Proceedings in Life Sciences

Perspectives in Chemoreception and Behavior

Edited by
R.F. Chapman, E.A. Bernays,
and J.G. Stoffolano, Jr.

With Contributions by
E.A. BERNAYS, L.M. BEIDLER, R.F. CHAPMAN,
A. GELPERIN, F.E. HANSON, T. JERMY, J.S. KENNEDY,
C. PFAFFMANN, M. ROTHSCHILD, D. SCHNEIDER,
L.M. SCHOONHOVEN, E. STELLAR

With 73 Figures, 3 in Full Color

Springer-Verlag
New York Berlin Heidelberg
London Paris Tokyo

R.F. Chapman
  Division of Biological Control and
  Department of Entomological Sciences,
  University of California,
  Berkeley, California 94706, U.S.A.

E.A. Bernays
  Division of Biological Control and
  Department of Entomological Sciences,
  University of California,
  Berkeley, California 94706, U.S.A.

J.G. Stoffolano, Jr.
  Department of Entomology
  University of Massachusetts
  Amherst, Massachusetts 01003, U.S.A.

The cover illustration shows the walking behavior of a fly in presence of two parallel lines of sugar solution. The thick line indicates a more concentrated solution. The solid leg shows that the fly makes contact with one line while feeding from the other.
Reproduced by persmission of V.G. Dethier, The Hungry Fly, Harvard University Press, © 1976.

Library of Congress Cataloging in Publication Data
Perspectives in chemoreception and behavior.
    (Proceedings in life sciences)
    Papers presented at a symposium held at the
University of Massachusetts, Amherst in May 1985.
    Bibliography: p.
    Includes index.
    1. Animals—Food—Congresses. 2. Chemoreceptors—
Congresses. 3. Insect–plant relationships—Congresses.
I. Chapman, R.F. (Reginald Frederick) II. Bernays,
E.A. (Elizabeth A.) III. Stoffolano, J.G.
IV. Series.
QL756.5.P47   1986   591.53   86-11811

Typeset by David E. Seham Associates Inc., Metuchen, New Jersey
Printed and bound by Quinn-Woodbine, Woodbine, New Jersey
Printed in the United States of America.

9 8 7 6 5 4 3 2 1

ISBN 0-387-96374-X Springer-Verlag New York Berlin Heidelberg
ISBN 3-540-96374-X Springer-Verlag Berlin Heidelberg New York

Vincent Gaston Dethier
A seventieth birthday tribute

Preface

In the study of the physiological basis of animal behavior Vince Dethier has been a pioneer, a guiding star. Although his own work has centered on the blowfly and the caterpillar, his interests and influence have spread far beyond the insects. The breadth of this impact is indicated by the contributions from colleagues and former students in this volume. These papers were originally presented at a meeting to honor Vince's 70th birthday held at the University of Massachusetts, Amherst, in May 1985. It was attended by friends and colleagues of all ages from many parts of the world.

However, the picture presented by these papers is not the whole story. What it does not show is the extent of Vince's interest and influence beyond the rigorous, though friendly, atmosphere of the research laboratory. His idyllic summers in Maine have produced studies on the natural history of feeding by insects culminating in *The Tent Makers*, with more to come. In these studies we see his real love and, dare we say, understanding of the insect.

Vince Dethier is not concerned simply with reaching the established scientist. In *To Know a Fly* he reaches out to those just beginning, perhaps even to those who will never begin, and provides insight both to the experimentalist's approach and to the fun of research. His sense of fun and his elegant, fluent writing have given us, too, his tongue-in-cheek fictional writings for children of all ages. This good humor and quiet, well-balanced view of things have been appreciated in more than his immediate scientific sphere of influence, and he has been and continues to be called on to play a significant role in university affairs.

For all of this, your science, your humor, your humanity, we thank you, Vince. A happy 70th birthday.

The Symposium at the University of Massachusetts, Amherst, was made possible through the generous support of the University. It was organized by Dr. R. Prokopy, Dr. J.G. Stoffolano, and Dr. G.A. Wyse, with assistance from many others. We are indebted to them all for making the meeting and, ultimately, this book possible.

We express our thanks also to Springer-Verlag for their cooperation in producing the volume.

Berkeley   R.F. Chapman
Berkeley   E.A. Bernays
Amherst   J.G. Stoffolano, Jr.

Contents

Contributors

The following is a list of contributors. The author's complete address is found on the first page of his or her contribution. Numbers in parentheses indicate the page on which the contribution begins.

E.A. Bernays (159)
L.M. Beidler (47)
R.F. Chapman (159)
A. Gelperin (33)

F.E. Hanson (99)
T. Jermy (143)
J.S. Kennedy (17)
C. Pfaffmann (59)

M. Rothschild (175)
D. Schneider (123)
L.M. Schoonhoven (69)
E. Stellar (1)

Chapter 1

The Internal Environment and Appetitive Measures of Taste Function in the Rat

ELIOT STELLAR*

It is the fact that animals are responsive to certain patterns of external environmental stimuli only when specific states of the internal environment prevail that specify certain basic biological motivated behaviors. On the biological side, we specify as a necessary condition the state of the internal environment, for example, in terms of peripheral changes such as the level of sex hormones or the condition of dehydration or salt depletion. In many of these cases, we have learned that these internal environment changes lead to changes in brain states which, in turn, yield the motivated behavior.

On the behavioral side, we measure motivated behavior in terms of the intensity or magnitude of the animal's response to the appropriate stimuli, such as the vigor of mating behavior or the rate or amount of eating by the food-deprived animal. Since eating changes the animal as the meal progresses, we have turned to studies of the approach to food stimuli that the animal only tastes. This appetitive measure can be repeated many times over short periods and is an ideal way to study an animal's responsiveness to taste stimuli as its internal environment and thus its brain states are varied experimentally.

Although Vince Dethier and I have agreed to disagree with each other over the question of whether the concept of motivation is useful in the study of taste and feeding behavior (Dethier 1982; Stellar and Stellar 1985), I want to acknowledge my debt to him, for his cogent arguments have led me to clarify my own thinking and to design experiments using appetitive as well as consummatory measures of behavior. I want to describe those experiments here because I think the appetitive measures we have used have enabled us to analyze, in new ways, the contribution of taste and other orosensory stimuli to the genesis and control of motivated behavior.

---

*David Mahoney Institute of Neurological Sciences, University of Pennsylvania, Philadelphia, Pennsylvania 19104, U.S.A.

Runway Method

Our experiments have been performed on rats, and we have chosen as our appetitive measure the speed of running to various taste stimuli in a meter-long runway (Figure 1.1). The animal is put in the start box and allowed to taste the stimulus being tested. When the rat turns and faces the door, the door is opened, starting a clock and allowing the rat the opportunity to run down the runway to the goal box where it puts its head in the hole, stops the clock, and reaches a reward of the same taste stimulus. Then the rat turns and runs in the opposite direction when the door is raised. Only 0.1 ml of taste reward or 0.1 g of food reward is given in each trial. Eleven trials are given for each determination, and the result expressed as the median of the last seven trials. The reciprocal of running time is taken to give running speed in cm sec^{-1}.

Under high deprivation conditions, the rat runs directly to the goal box and the taste reward, leaving the start box like a track star at the starting line. Under low deprivation, when the rat is satiated, or under low reward conditions, it moseys out of the start box, runs a short distance, stops and explores the runway, sniffs, grooms, and scratches before proceeding to the goal box where it may stand in front of the hole giving access to the reward cup before actually putting its head in and tasting the reward. Thus the measure of motivated behavior in the runway is not just how fast the animal runs, but is a measure of the competition among drives: running directly to food or water versus grooming, exploring, sniffing, scratching.

Since the rewards are so small, the rats get about a milliliter of fluid or a gram of food in the course of the 11-trial test. Thus, postingestional effects are kept at a minimum, and the test is one of the responsiveness to taste and other orosensory factors under specified conditions of the internal environment or of actual brain states.

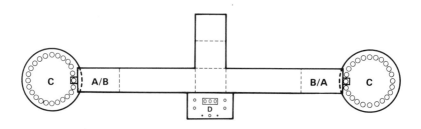

2-way runway

Figure 1.1. Runway used to measure appetitive motivation (1 m long). A. Start-box. B. Goal-box. C. Drinking tray. D. Timer. (Reprinted with permission from Zhang DM, Stellar E, Epstein AN. Together intracranial angiotensin and systemic mineralcorticoid produce avidity for salt in the rat. Physiol Behav 32 ©1984, Pergamon Press, Ltd.)

Role of the Internal Environment

How does the change in the internal environment change the animal's response to taste and other orosensory factors? The answer, we believe, is in the effect of the internal environment on the brain, either over afferent neural pathways or through more direct effects of hormones (peptides and steroids) or other changes (e.g., temperature, osmotic pressure) on specified target neurons. Some examples follow.

Thirst Motivation

We know that water deprivation produces both osmotic and volumetric changes in the internal environment (Epstein et al. 1973). The volumetric changes are mediated by the renin-angiotensin cycle, and it has been shown that angiotensin is a potent dipsogen (Epstein et al. 1970). In our runway experiment (Zhang et al. 1984), we reported that running speed is an increasing, monotonic function of hours of water deprivation that parallels the water intake function (Figure 1.2). Using rats that had *not* been water-deprived, we were then able to show that injection of angiotensin II into the lateral and third ventricles caused them to run avidly for small rewards of water (Figure 1.3). The higher the dose of angiotensin, the faster they ran and the longer the effect persisted over a 2-hr period of repeated testing. The remarkable thing is that although the half-life of angiotensin is measured in minutes, its effect on running to water lasted for an hour or longer.

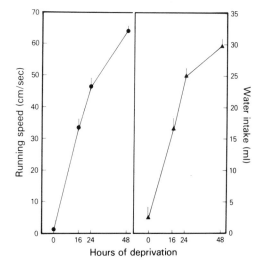

Figure 1.2. Running speed (left) and water intake (right) as a function of hours of water deprivation. (Reprinted with permission from Zhang DM, Stellar E, Epstein AN. Together intracranial angiotensin and systemic mineralcorticoid produce avidity for salt in the rat. Physiol Behav 32 ©1984, Pergamon Press, Ltd.)

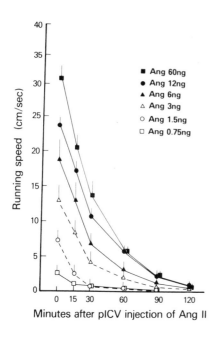

Figure 1.3. Running speed of undeprived rats as a function of pulse intracerebroventricular (pICV) injection of different doses of angiotensin II (Ang), measured up to 2 hr after injection. (Reprinted with permission from Zhang DM, Stellar E, Epstein AN. Together intracranial angiotensin and systemic mineralcorticoid produce avidity for salt in the rat. Physiol Behav 32 ©1984, Pergamon Press, Ltd.)

Salt Hunger

We also found that when intraventricular angiotensin was combined with a subcutaneous dose of desoxycorticosterone acetate (DOCA), a precursor of the hormone aldosterone, the animals also ran fast to 3% NaCl solutions (Figure 1.4, right). Because of this, Jay Schulkin, Phil Arnell, and I decided to look more closely at the motivation for salt. The striking thing about this fast running to 3% NaCl is the fact that the normal rat has a negative response to 3% NaCl (over 0.5 M) as judged by his unwillingness to drink it. In fact, our previous work (Stellar et al. 1954) had shown that intake of NaCl solutions is a rising

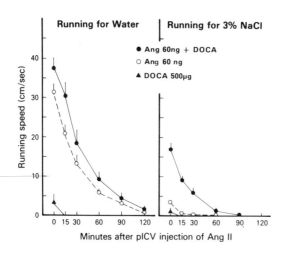

Figure 1.4. Speed of running to water (left) and 3% NaCl (right) as a function of treatments with desoxycorticosterone acetate (DOCA) and angiotensin (Ang). (Reprinted with permission from Zhang DM, Stellar E, Epstein AN. Together intracranial angiotensin and systemic mineralcorticoid produce avidity for salt in the rat. Physiol Behav 32 ©1984, Pergamon Press, Ltd.)

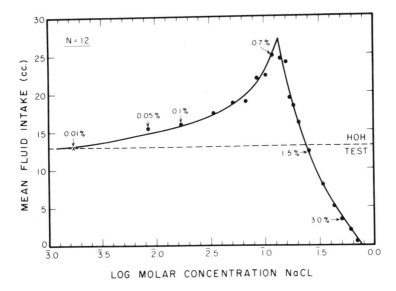

Figure 1.5. Preference-aversion function for NaCl solutions of increasing concentration, measured by the amount ingested in a 1-hr. test. (Reprinted with permission from Stellar ES, Hyman R, Samet S. Gastric factors controlling water- and salt-solution drinking. J Comp Physiol Psychol 47 © 1954, American Psychological Association.)

and falling function of concentration, referred to as a preference-aversion function (Figure 1.5). Making rats salt-deficient by adrenalectomy (Epstein and Stellar 1955) greatly increases their ingestion of all NaCl concentrations, but there is still a relative aversion to 3% NaCl solutions (Figure 1.6, right).

Schulkin et al. (1985) report that in the runway test, normal rats show a steady decline in running speed to increasing NaCl concentrations (Figure 1.7, top), in spite of the fact that, at the same time, they show the rising-falling preference-aversion functions in intake tests (Figure 1.7, bottom). So the rising portion of the preference-aversion curve is not necessarily increasing response to increased intensity of tastes in the normal rat. However, if the same rats are made salt-hungry by putting them on a salt-deficient diet and giving them large (10 mg) doses of DOCA subcutaneously for 2 days, they run very rapidly to all concentrations of NaCl, including not only 3%, but also 6% (Figure 1.7, top, dashed line). Even more remarkable is the fact that in a repetition of this experiment, water-satiated rats ran rapidly for NaCl concentrations as high as 24% (over 4.0 M) (Figure 1.8). Some of them even ran for solid salt crystals, reminiscent of the drive toward a salt lick seen in deer and cattle in nature.

In the same experiment, we also found that rats ran to other solutions that taste salty to humans (e.g., sodium bicarbonate and lithium chloride). Furthermore, if we gave the rats a chance to drink 5.0 ml of 3% NaCl one-half hour or an hour before going into the runway, they did not run rapidly to 3% NaCl, evidence for a change in the internal environment amounting to a satiating effect.

Epstein (1982) has suggested that the critical factor in salt hunger in the rat is not the level of sodium in the internal environment, but rather the changes

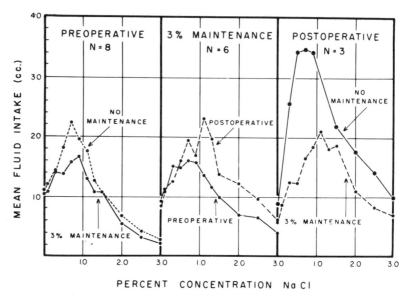

Figure 1.6. Effects of adrenalectomy on the preference-aversion intake functions for NaCl. The right panel shows the dramatic effect of adrenalectomy in salt-depleted rats (no maintenance) compared to rats with access to 3% NaCl 5 hr/day. The middle panel shows that normal and adrenalectomized rats are similar when both are maintained on 3% salt. The left panel shows that salt maintenance makes little difference in the normal rat. (Reprinted with permission from Epstein AM, Stellar E. The control of salt preference in the adrenalectomized rat. J Comp Physiol Psychol 48 © 1955, American Psychological Association.)

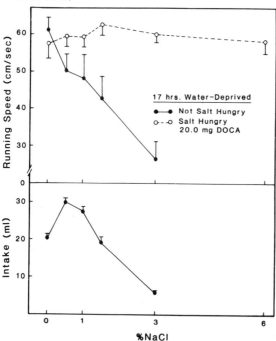

Figure 1.7. Speed of running to different concentrations of NaCl in normal rats (top) compared with preference-aversion function (bottom). The dashed line in the top panel shows the running speed when the animals are made salt-hungry. (Reprinted with permission from Schulkin J, Arnell P, Stellar E. Running to the taste of salt in mineralcorticoid-treated rats. Horm Behav 19 © 1985, Academic Press.)

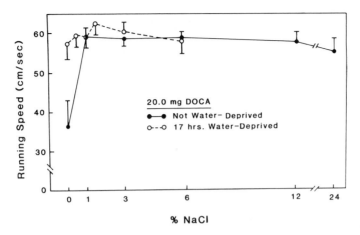

Figure 1.8. Speed of running of salt-hungry rats to concentrations of NaCl as high as 24% (solid line). (Reprinted with permission from Schulkin J, Arnell P, Stellar E. Running to the taste of salt in mineralcorticoid-treated rats. Horm Behav 19 © 1985, Academic Press.)

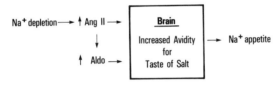

Figure 1.9. Epstein's synergy hypothesis of salt hunger, showing how angiotensin and aldosterone act together in the brain to produce the salt appetite. (Reprinted with permission from Zhang DM, Stellar E, Epstein AN. Together intracranial angiotensin and systemic mineralcorticoid produce avidity for salt in the rat. Physiol Behav 32 ©1984, Pergamon Press, Ltd.)

in hormones that the sodium-deficient state produces. According to this view (Zhang et al. 1984), the hormones of salt conservation at the kidney, angiotensin and aldosterone, act in synergy in the brain to yield the brain state responsible for salt appetite (Figure 1.9). Preliminary experiments in Epstein's laboratory now show that blocking the synthesis of active angiotensin and blocking aldosterone receptors in the brain can block salt appetite even in the most severe salt depletion (personal communication). Thus in the case of salt hunger, an internal environment state (salt deprivation leading to angiotensin and aldosterone increases) acts on the brain to yield increased responsiveness to the taste of salt stimuli, leading to increased ingestion of NaCl and the correction of the initiating condition of salt depletion.

Cholecystokinin (CCK) and Satiation

Whereas angiotensin arouses thirst and salt motivation, CCK is a putative satiety hormone, contributing toward a reduction of food intake in many mammals, including rats, monkeys, and humans. CCK is secreted by intestinal cells when food, particularly fats and proteins, enters the stomach. In the first experimental studies, Gibbs et al. (1973) injected CCK intraperitoneally and produced a marked reduction in the food intake of rats. Later, these same investigators showed that peripherally injected CCK lost its effect if the vagus nerve was cut (Smith et al. 1981), showing that its route to the brain was over an afferent neural pathway.

By this time, however, it was already known that CCK was synthesized in the brain and that there were receptors for it in widespread regions, from the cortex to the area postrema in the caudal medulla (Zarbin et al. 1983). Furthermore, Della-Fera and Baile (1979) had shown that, in sheep, intraventricular infusion of the octapeptide CCK-8 produced marked reduction in food intake. More than that, in a subsequent study, they showed that antibodies to CCK-8 resulted in increased food intake when infused intraventricularly in sheep (Della-Fera et al. 1981). They had difficulty, however, in replicating these findings in the rat, as did other investigators (Grinker et al. 1980). Maddison (1977), on the other hand, reported positive findings with intraventricular injection of CCK-8 in rats and reported reductions in lever-pressing for food.

We followed Maddison's lead and examined the effects of CCK-8 injected into the lateral and third ventricles of the rat on both running speed in the runway and in a 1-hr test of food intake in the home cage. Dian-Ming Zhang, Wlodymur Bula, and I (1986) first demonstrated that intraperitoneal injections of CCK-8 ($8\mu g\ kg^{-1}$) reduced speed of running to wet mash food rewards when tested 0, 15, 30, and 60 min after injection (Figure 1.10); it also reduced their food intake (Figure 1.11). Isotonic saline was used as the control injection in these tests, and each animal served as its own control in a 2-day, crossover design.

Cannulas were then implanted stereotaxically into the lateral and third ventricles of these rats. Tests with intraventricular angiotensin produced prompt drinking in the home cage, demonstrating our successful access to the ventricles. Furthermore, histology and terminal tests with the injection of India ink into

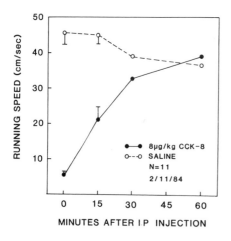

Figure 1.10. Effect of intraperitoneal (IP) injection of cholecystokinin octapeptide (CCK-8) in suppressing speed of running to food, measured over a 1-hr. period. (Reprinted with permission from Zhang DM, Bula W, Stellar E. Brain cholecystokinin as a satiety peptide. Physiol Behav 36 ©1986, Pergamon Press, Ltd.)

Figure 1.11. Suppressing effect of intraperitoneal (IP) injection of CCK-8 on food intake in a 1-hr. test. (Reprinted with permission from Zhang DM, Bula W, Stellar E. Brain cholecystokinin as a satiety peptide. Physiol Behav 36 ©1986, Pergamon Press, Ltd.)

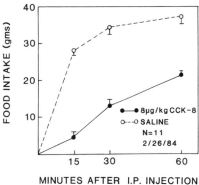

the ventricles showed that our placements were such that the injections reached all regions of the ventricular system (Figure 1.12).

Thus prepared, the rats were then given pulse intracerebroventricular (pICV) injections of 1/100th the intraperitoneal dose of CCK-8 or 0.08 µg kg^{-1} (80 ng kg^{-1}), immediately before being placed in the runway. The 11-trial tests were repeated at 15, 30, and 60 min after the pICV injection. Desulfated CCK-8, which is almost inactive, was injected as the control in a 2-day, crossover design in which each rat again served as its own control. As Figure 1.13 shows, the 0.08 µg dose of CCK produced a marked and prolonged depression in speed of running to food rewards (left panel). Cutting the dose in half to 0.04 µg lost the effect (third panel) and a 0.06 µg kg^{-1} dose produced an intermediate and

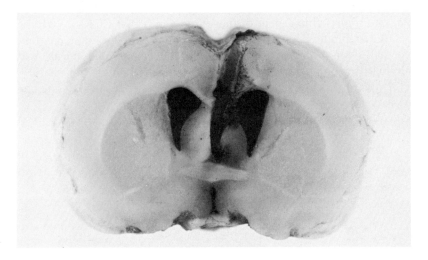

Figure 1.12. Coronal section through the rat's brain, showing how terminally injected India ink enters the lateral and third ventricles. (Reprinted with permission from Zhang DM, Bula W, Stellar E. Brain cholecystokinin as a satiety peptide. Physiol Behav 36 ©1986, Pergamon Press, Ltd.)

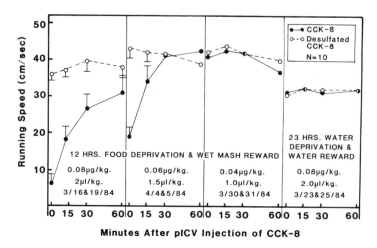

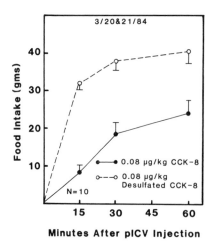

Figure 1.13. Suppressing effect of pulse intracerebroventricular (pICV) injection of CCK-8 on speed of running to food reward, as a function of dose (three left panels). Control run to water reward (shown on right). (Reprinted with permission from Zhang DM, Bula W, Stellar E. Brain cholecystokinin as a satiety peptide. Physiol Behav 36 ©1986, Pergamon Press, Ltd.)

short-lived depression of speed of running to food. That the depression of running speed was specific to hunger motivation is shown in the right panel where it is seen that 0.08 μg CCK-8 had no effect on speed of running to water rewards in the thirsty rat. Similar effects of pICV-8 of CCK-8 were obtained in the food intake tests (Figure 1.14) where a marked reduction of ingestion was evident.

Again, one remarkable aspect of these findings is that CCK-8, with a half-life measured in minutes, has effects on running speed and food intake that last up to one-half hour and perhaps longer. As in the case of angiotensin, we can conclude that the effects of CCK are both peripheral and central neural. The peripheral effects are conducted to the brain over the vagus nerve. The central

Figure 1.14. Suppressing effect of pulse intracerebroventricular (pICV) injection of CCK-8 on food intake. (Reprinted with permission from Zhang DM, Bula W, Stellar E. Brain cholecystokinin as a satiety peptide Physiol Behav 36 ©1986, Pergamon Press, Ltd.)

effects may involve direct action of CCK on the brain. Since CCK does not readily penetrate the blood–brain barrier, two possibilities suggest themselves. Either the conditions of satiation trigger CCK synthesis and secretion in the brain, in parallel with the triggering of synthesis and secretion in the intestine, so that it reaches pertinent target organs within the brain, or the peripheral CCK enters the brain through one of the circumventricular organs which have little or no blood–brain barrier (Figure 1.15).

In the light of these possibilities, the area postrema in the caudal medulla, one of the seven known circumventricular organs, suggests itself. One reason is that many of the terminals of the vagus nerve reach the area postrema and the adjacent caudal medial nucleus of the solitary tract (Hyde and Miselis 1983). Furthermore, van der Kooy (1984) has recently shown that lesions of the area postrema and surrounding structures block the satiety effects of intraperitoneal CCK-8 injections in the rat.

Since the flow of the cerebrospinal fluid is posterior and since the area postrema has a minimal cerebrospinal fluid–brain barrier, we were encouraged to think that perhaps our injections of CCK-8 into the lateral and third ventricles were having their effects in this region of the caudal medulla. Therefore, working with Dr. William Flynn, we tested rats in the runway that were prepared with both lateral and third ventricular cannulas and with fourth ventricular cannulas. The lateral and third ventricular placements were verified with the angiotensin tests of drinking. The fourth ventricular placements were verified by Dr. Flynn by the hyperglycemia produced by injections of 5-thio-glucose (5TG) there.

Figure 1.16 shows the results of testing rats with the lateral and third ventricle placement and with the fourth ventricle placement, injecting 0.08 μg kg^{-1} and 0.12 μg kg^{-1} in two tests where the control consisted of the same doses of desulfated CCK-8 in the 2-day, crossover design. Since the results were virtually

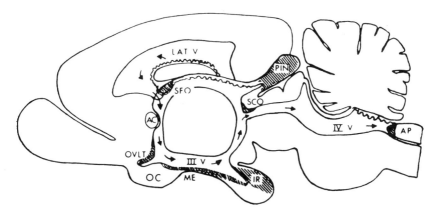

Figure 1.15. Lateral view of the rat's brain, showing the seven circumventricular organs that have little or no blood–brain barrier. SFO subfornical organ; OVLT, organum vasculosum of the lamina terminalis; ME median eminence; IR infundibular recess and neurohypophysis; SCO subcommissural organ; PIN pineal gland; AP area postrema. (Reprinted with permission from Phillips MI. Neuroendocrinology, Vol 25: Angiotensin in the Brain. © 1978 S. Karger AG, Basel.)

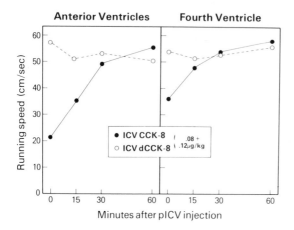

Figure 1.16. Suppression of running speed by pulse intracerebroventricular (pICV) injection of CCK-8 into the lateral and third ventricles (left) and the fourth ventricle (right). Control injections were made with desulfated CCK-8 (dCCK-8).

the same in the two tests, the values were combined into a single figure. Quite clearly, these preliminary results show that, contrary to our hypothesis, CCK-8 injection into the fourth ventricle is less effective in suppressing running speed than injection into the lateral and third ventricles. Whether this is because the injected CCK exits from the fourth ventricular foramina into the subarachnoid

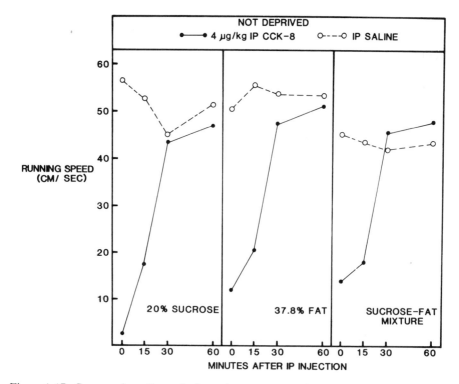

Figure 1.17. Suppression of speed of running to sucrose, fat, and a sucrose-fat mixture by the satiated rat following intraperitoneal (IP) injection of CCK-8.

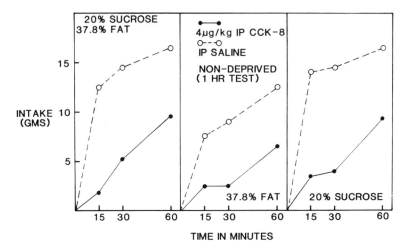

Figure 1.18. Suppression of ingestion of sucrose, fat, and a sucrose-fat mixture by the satiated rat following intraperitoneal (IP) injection of CCK-8.

space and thus does not make as full contact with CCK receptor sites as the flow from the more anterior ventricular injections, we do not know. To test this possibility, we will try slow infusion into the fourth ventricle rather than pulse injection. Making lesions in the region of the area postrema before intraventricular injections is another approach we plan to take. In addition, we want eventually to investigate the role of the paraventricular nucleus (PVN) in the walls of the anterior third ventricle as a possible site for CCK's satiety effects. The PVN is known as a part of the satiety mechanism (Kapatos and Gold 1973; Hoebel 1984).

Before closing, I should mention that one of my students, Howard Weiner, and I have also been studying the peripheral and central effects of CCK on the speed of the rat's running to sugar and fat rewards. Here the rewards are so potent that the animals will run without having been deprived of food. Thus we may be examining the effects of CCK on pure taste (sucrose) and orosensory (fat) responsiveness, apart from the internal environment and brain state changes that accompany food deprivation. Figures 1.17 and 1.18 show our preliminary findings on running speed and ingestion, following intraperitoneal CCK-8 injection, and Figure 1.19 shows the results following intracerebroventricular injection of CCK-8. Thus it may be that the putative satiety hormone CCK also diminishes the effectiveness of taste and other orosensory rewards. This idea fits with motivational theory (Stellar and Stellar 1985) which holds that taste and other orosensory stimuli act as incentives that arouse drive and motivation at the same time that they function as rewards in the learning and performance of appetitive behaviors such as the running to food, salt, water, etc. in the runway. Sweet, painful, and bitter stimuli, for example, arouse motivated behavior without depending on changes in the internal environment. So we have to broaden our definition of basic biological motivation to include the direct

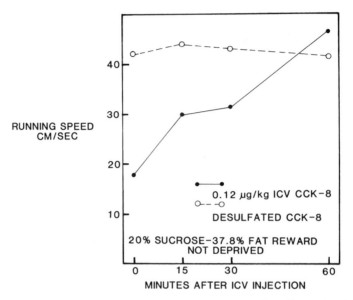

Figure 1.19. Suppression of speed of running to a sucrose-fat mixture by the satiated rat following pulse intracerebroventricular (ICV) injection of CCK-8.

impact of taste and other sensory stimuli as well as the effects of changes in the internal environment on the animal's responsiveness.

Conclusions

The results of our experiments with rats tested in both the runway (appetitive behavior) and the home-cage ingestion tests (consummatory behavior) suggest a number of conclusions.

1. Running speed is a valuable measure of motivated behavior that parallels the ingestive behavior measure, but is free of the postingestional effects of ingestion, so that it may be used as a test of taste and orosensory responsiveness.
2. The internal environment is important in both the arousal (e.g., angiotensin) and satiation (e.g., cholecystokinin) of motivated behavior, in both runway and ingestion tests.
3. Changes in the internal environment produce their effects by altering brain states involved in motivated behavior, either by way of afferent neural pathways (e.g., vagus) or by more direct effects on the brain (e.g., via the circumventricular organs or by triggering hormone synthesis in the brain).
4. The changes in the internal environment and in the brain responsible for the arousal and satiation of motivated behavior are also responsible for the changes in taste-responsiveness and the effectiveness of taste and orosensory rewards, measured by running speed in the runway.

Acknowledgement This work was supported by a grant from the Whitehall Foundation.

References

Della-Fera MA, Baile CA (1979) Cholecystokinin octapeptide: continuous picomole injections into the cerebral ventricles of sheep suppress feeding. Science 206:471–473

Della-Fera MA, Baile CA, Schneider BS, Grinkler JA (1981) Cholecystokinin antibody injected in cerebral ventricles stimulates feeding in sheep. Science 212:687–689

Dethier VG (1982) The contribution of insects to the study of motivation. In: Morrison A, Strick P (eds) Changing Concepts of the Nervous System. Academic Press, New York, pp 445–455

Epstein AN (1982) Mineralcorticoids and cerebral angiotensin may act together to produce sodium appetite. Peptides 3:493–494

Epstein AN, Stellar E (1955) The control of salt preference in the adrenalectomized rat. J Comp Physiol Psychol 48:167–172

Epstein AN, Fitzsimons JT, Simons BJ (1970) Drinking induced by injection of angiotensin into the brain of the rat. J Physiol Lond 210:457–475

Epstein AN, Kissileff HR, Stellar E (1973) The neuropsychology of thirst: new findings and advances in concepts. VH Winston and Sons, Washington, DC, pp 37–112

Gibbs J, Young RC, Smith GP (1973) Cholecystokinin decreases food intake in rats. J Comp Physiol Psychol 84:488–495

Grinker JA, Schneider BS, Ball G, Cohen A, Strohmayer A, Hirsch J (1980) Cholecystokinin (CCK-8) and bombesin (BBS) intracranial injections and satiety in rats. Proc Fed Am Soc Exp Biol 30:501

Hoebel BG (1984) Neurotransmitters in control of feeding and its reward: monoamine, opiates, and brain-gut peptides. In: Stunkard AJ, Stellar E (eds) Eating and Its Disorders. Raven Press, New York, pp 15–38

Hyde TM, Miselis RR (1983) Effects of area postrema/caudal medial nucleus of the solitary tract lesions on food intake and body weight. Am J Physiol 244:577–587

Kapatos G, Gold RM (1973) Evidence for ascending noradrenergic mediation of hypothalamic hyperphagia. Pharmacol Biochem Behav 1:81–87

Maddison S (1977) Intraperitoneal and intracranial cholecystokinin depress operant responding for food. Physiol Behav 19:819–824

Schulkin J, Arnell P, Stellar E (1985) Running to the taste of salt in mineralocorticoid treated rats. Horm Behav 19:413–425

Smith GP, Jerome C, Cushin HJ, Eterno R, Simanski KJ (1981) Abdominal vagotomy blocks the satiety effect of cholecystokinin in the rat. Science 213:1036–1037

Stellar JR, Stellar E (1985) The Neurobiology of Motivation and Reward. Springer-Verlag, New York

Stellar ES, Hyman R, Samet S (1954) Gastric factors controlling water- and salt-solution-drinking. J Comp Physiol Psychol 47:220–226

van der Kooy D (1984) Area postrema: site where cholecystokinin acts to decrease food intake. Brain Res 295:345–347

Zarbin MA, Innis RB, Wamsley JK, Snyder SH, Kuhar MJ (1983) Autoradiographic localization of cholecystokinin receptors in rodent brain. J Neurosci 3:877–906

Zhang DM, Bula W, Stellar E (1986) Brain cholecystokinin as a satiety peptide. Physiol Behav 36:1183–1186

Zhang DM, Stellar E, Epstein AN (1984) Together intracranial angiotensin and systemic mineralocorticoid produce avidity for salt in the rat. Physiol Behav 32:677–681

Chapter 2

Animal Motivation: The Beginning of the End?

John S. Kennedy*

This chapter is a salute to Vince Dethier as a liberator in the realm of behavioral theory. We have his own word for it that he was motivated by much more than every scientist's natural curiosity through all those years of masterly physiological analysis of a fly's feeding behavior:

> The purpose underlying the foregoing analysis of the fly was not to study the fly *sui generis*. It was to hold a different glass to the problem of motivation, to ascertain whether or not it occurred in an organism whose evolutionary appearance predated that of man, to discover whether or not its expression lay within the capabilities of a relatively simple nervous system, to enquire into the reality of motivated behavior as a separate class. (Dethier 1966, p 128)

Snark Hunting

He started out assuming that some animals really are motivated:

> In the higher organism . . . the motivational aspects of instinctive behavior emerge clearly, and it is quite meaningful to talk about 'drives', 'goal-directed behaviour,' and 'satiation.' (Dethier and Stellar 1961, p 80)

There were, however, dyed-in-the-wool Behaviorist opponents of such anthropomorphic terms still around, notably T.C. Schneirla, claiming motivation was an unnecessary and misleading assumption (see Dethier 1964, 1966). Dethier proceeded, as no one else has, to give their claims a full and fair test by meticulous and exhaustive experiments to see how far a fly's feeding behavior could be explained in familiar physiological, input-output terms, without invoking motivation. At some point this endeavor could fail and the unexplained residue would *be* motivation, isolated and identifiable at last.

---

*Department of Zoology, University of Oxford, South Parks Road, Oxford OX1 3PS, U.K.

But that point never came: the Snark was a Boojum. The upshot of his un-remitting quest was that no unexplained residue could be found in the fly. More than that: he could not pinpoint any such residue in accounts of feeding behavior in the rat, either, although the rat is a much "higher" animal and its feeding behavior has been a standard model of motivated behavior (Dethier 1966, 1976, 1981, 1982).

The issue at stake here was not really what motivation is. There is no con-sensus on that as testified most recently by Halliday (1983, p 101): the meanings attached to it range all the way from near-supernatural to strictly mechanistic. Most contemporary zoologists take the mechanistic view:

> At best, intervening variables are labels for unknown physiological processes . . .
> The ultimate aim of much research into motivation is to identify and understand
> how such processes work, so that concepts such as hunger, thirst and drive become
> unnecessary. (Halliday 1983, p 105)

Until that day dawns they make free use of terms embodying these concepts borrowed from psychology, along with many others including motivation (not to mention mind-reading, manipulation, and rape!) that imply a conscious mind. Nevertheless, with few exceptions (Thorpe 1954, 1963; Collier 1980; Baker 1982; Griffin 1984) they disclaim the mentalist connotation of these terms when applied to animals, saying (e.g., Krebs and Davies 1981, p 256) they are used only as a convenient shorthand way of describing the adaptive *functions* of behavior: its evolutionary or "ultimate" causes, not its immediate or "proximate" ones.

Although these two kinds of cause are quite distinct, ultimate causes are easily mistaken for proximate ones when the behavior is described anthropo-morphically in mentalist, teleological terms such as motivation (Tinbergen 1951, pp 4–5; Kennedy 1967, 1972, 1985; Hinde 1970, p 417; Krebs and Davies 1981, p 256). The reason is simple. The adaptive function of any action we take (when it is adaptive!)—what we gain from it in fitness—is its ultimate cause, as in animals. Subjectively, however, its function is also its purpose, conceived as a *proximate* cause of it.

Nevertheless the admitted risk of confusing those two kinds of cause is gen-erally ignored even by scientists because mentalist language is so convenient. It is the language we learn early to describe the behavior we know the best, our own, so we have a rich, instant vocabulary for it (Kennedy 1986b). It trips off the tongue, making for easy, adaptable discourse about the behavior of animals too. It comes naturally to us to describe behavior as if it were purposeful, simply because in ourselves it is. So we all use such language to describe animal behavior and assuredly will go on doing so.

In addition, this teleological way of thinking is genuinely useful in research inasmuch as it identifies what the animal gains from any given behavior. We all want to have some idea of that even when we are studying only the proximate causes of behavior. It suggests what known kinds of mechanism could be em-ployed in such a task, and they can then be looked for. However, this *indirectly* heuristic value of mentalist language is entirely beside the point when it comes to characterizing motivation as a separate class of proximate cause of behavior. This is exceedingly difficult because, as Dethier (1981, 1982) documented at length, the concept of motivation remains fuzzy, elusive, and controversial, in his words. So, from the many criteria found in the literature—varying responsiveness, endogenous activity ("the essence of motivation": Dethier 1966, p 120), drive, drive reduction (satiation), goal-directedness, etc.—Dethier carefully assembled an inclusive definition. To proceed with investigating the physiological mechanism, he of course had first to translate the psychological terms into physiological-behavioral equivalents:

> Motivation is a specific state of endogenous activity in the brain which, under the modifying influence of internal conditions and sensory input, leads to behavior resulting in sensory feedback or change in internal milieu, which then causes a change (reduction, inhibition, or another) in the initial endogenous activity. (Dethier 1964, p 1139)

But the quest for motivation thus objectively defined drew a blank, since this definition "would construe all behavior as motivated" (Dethier 1964, p 1140). Motivation then became a superfluous concept: "Certainly the concept adds nothing to our understanding of the fly's feeding behavior" (Dethier 1976, p 474).

Unnecessary Impediment

Undismayed, Dethier turned hopefully to a more rigorous criterion, operant conditioning, for an acid test. The idea was that any behavior that could be so conditioned would be motivated behavior because, it was argued,

> The operant is essentially a voluntary act, not dependent upon specific afferent input, that an animal can use to obtain reinforcement. Since the animal exerts control over the occurrence of its response, the behavior is distinct from reflexes and from complex fixed motor patterns. (Dethier 1964, p 1139; 1966, p 118)

He did not succeed in conditioning his flies operantly but was later able to cite at least one report of success with an insect (1981, p 464). To my mind, be it said, the above argument that success would demonstrate motivation is unconvincing. During the conditioning process the animal performs a variety of actions, many of them dependent on specific afferent inputs, and the experimenter chooses to reinforce one of them. Thus it is not the animal but the experimenter who controls which action gets conditioned; it is the experimenter's motivation, not the animal's, that increases the frequency of that particular action by the animal. There is no sign here of a special motivating component in the animal's behavior. An alternative, and more plausible, interpretation would be that motivation simply means learning (Dethier 1981, p 465); but then the term motivation is again redundant.

For whatever reason, Dethier did not pursue the argument from operant conditioning to motivation after 1966; but for the time being he determinedly kept an open mind on the major question of mental activity in the fly despite his lack of any positive evidence of it so far, and defended his continuing quest:

> We might do well to accept in principle the dualistic methodology . . . namely, to use, in addition to behavioristic and physiological analyses, concepts about phys- iological events which come to us through common sense, intuition, introspection, sensation, and perception . . . Perhaps these insects are little machines in a deep sleep, but looking at their rigidly armored bodies, their staring eyes, and their mute performances, one cannot help at times wondering if there is anyone inside. (Dethier 1964, p 1145)

As the quest continued doggedly through the 1970s, however, he seems to have lost faith in this equivocal approach and eventually, after a close scholarly examination of existing ideas on the subject, reached this conclusion:

> Analysing the concept . . . suggests it may have outlived its usefulness . . . Perhaps motivation ought to be a poetical expression not to be taken seriously. (Dethier 1981, p 466)

This negative but gentle formulation left it open to everyone's personal judgment whether or not to use the term motivation. One could grant that it is not strictly necessary but nevertheless decide that, on balance, the practical convenience justifies using it and other mentalist terms—on the assumption that as mere metaphors they do negligible harm. Indeed, W.T. Keeton spoke for a great many people who see no clear dichotomy between reflexes and more complex behavior but still think it best to maintain a dichotomy of terms, when he wrote:

> There is no difference in kind between simple reflexes and more complex reactions . . . but . . . applying the term 'reflex' in such a broad manner makes it synonymous with 'behavior', which does not help us at all. It is customary, therefore, to restrict the term 'reflex' to relatively simple and automatic responses to stimuli and to designate more complicated behavior patterns by other terms. (Keeton 1967, p 455)

That sounds like plain common sense—until one notices that those other terms are, in fact, different in kind from reflexes since they are not objective but subjective, loaded with anthropomorphic overtones. At best they are simply metaphors for ultimate causes, whereas reflexes are proximate ones. Keeton thus unwittingly reintroduced the very dichotomy he had wittingly just dis- avowed. Dethier, having rejected that dichotomy, soon went considerably fur- ther than he had in 1981 and rejected also the conventional assumption that such metaphorical terms do negligible harm:

> The concept of motivation . . . has not only outlived its usefulness as an analytical scaffolding but has become an impediment to our understanding of the behavior it purports to explain. (Dethier 1982, p 454)

"An impediment": that is Dethier's unfashionable but valid and crucial point, which Keeton and most workers have missed.

Cartesian Dichotomy

How did the blind spot arise? Whatever the term motivation has been taken to mean, the underlying issue has never been that, but rather whether motivation and related endogenous events do constitute *a separate class* of causal mechanism in behavior, as Dethier put it (1966; see p 17). The idea that they do, goes back to long before Descartes as everyone knows, but he articulated it so clearly, in the language of his time, that it has come to be called Cartesian dualism (Dethier 1981, p 462). He restricted the idea to man. In our time it was Lorenz (1950) who most unequivocally extended the dualism to all animals. Internal "energy" accumulation and external stimulation, he wrote (p 251), are "two absolutely heterogeneous causal factors," and he insisted on

> the peculiarity and independence of endogenous activity as a distinct physiological process . . . an independent, particulate function of the central nervous system which . . . is, at the very least, equally as important as the reflex. (Lorenz 1950, p 249)

The dichotomy is usually expressed less starkly nowadays but it is often implied, as for example in Halliday and Slater's Introduction to their up-to-date, multiauthored textbook:

> As animals become more complex . . . the problem of changing motivation crops up. (Halliday and Slater 1983, p 3)

> While single actions may appear in much the same form every time . . . sequences of different activities are seldom repeated in exactly the same order . . . This is also a good reason why changing motivation often has to be invoked. (Halliday and Slater 1983, p 6)

If motivation has to be invoked only sometimes, contrary to Dethier's conclusion (1964; see p 19), then the unavoidable inference is that there exist simpler behaviors for which motivation does not have to be invoked. Motivation is inferred only when responsiveness changes and if that does not change then motivation itself is not invoked.

Nonmotivated Behavior

Obviously there would be no dichotomy without that idea of a *non*motivated class of behavior. So the idea deserves closer scrutiny than it usually receives in discussions of motivation including Dethier's. Turning attention to nonmotivated behavior has one substantial advantage. Whereas motivated behavior is multivalent and controversial, the meaning of nonmotivated behavior is clear and agreed on in the psychological and ethological literature. It is said to be made up of "simple reflexes" characterized by one property conveyed in many ways: "stimulus-bound," "push-button," "S-R" (stimulus-response), "automatic," "rigid," invariable," "inflexible," "stereotyped," "lacking endogenous neural activity,'' and so on. Thus Dethier and Stellar (1961, p 70) described nonmotivated behavior as "complex chaining of simple reflexes with invariable S-R relationships," and Dethier (1964, p 1142) once wrote of "reflex physiology with its assumption of neurological silence in the absence of overt stimulation" (which he understandably went on to say had "failed to provide a basis for explaining behavior even in insects").

The fact is that this picture of reflex action is a relic from the early decades of this century when militantly Mechanist zoologists overreacted to the previous Vitalist obscurantism and oversimplified understandably, but grotesquely (see Lorenz 1950; Tinbergen 1951; Kennedy 1958, 1966, 1967), in their determination to discard it. Their model of reflex action still survives as the antonym of motivated behavior, but it bears no resemblance to the real reflex action that the physiologist Sherrington (1947, reprinted from 1906) had already explored in depth and summed up in his famous title "The *Integrative* Action of the Nervous System" (my emphasis). He insisted that the "simple reflex" was an abstraction. So later did the insect physiologist Wigglesworth (1939, 1966); and the psychologist Staddon has recently recapitulated the ample evidence that

> Sherrington's concept of the reflex is far from the simple, inflexible, push-button caricature sometimes encountered in introductory textbooks. (Staddon 1983, p 30)

That is an understatement. Unfortunately the push-button caricature is by no means confined to introductory texts. It is all-pervasive, providing the only logic behind the ubiquitous use of "motivation" to mean some nonreflex mechanism. Staddon continues:

> To be sure, there is always a stimulus and a response; but the ability of the stimulus to produce the response depends on the reflex threshold—and the threshold of each reflex depends not only on the state of many other reflexes but also on higher centers, which retain the effects of an extensive past history . . . The function of reflexes is the *integration* of behavior, which would be impossible without well-defined rules of interaction. (Staddon 1983, pp 30–31)

Staddon has put the Mechanist (now called "radical") Behaviorists' historic mistake in a nutshell:

> Psychologists emphasised the wrong aspect of reflexology, attending to the stimulus-response property of reflexes, and not to reflex integration—the principles by which tendencies to action combine to produce overt behavior. (Staddon 1986, p 84)

Zoologists have done the same. Consequently "reflex action" seems to exclude most of behavior, and above all the making, storing, and using of spatio-temporal "maps," which is the integration of *conditioned* reflexes, although usually known as cognition. This is an example of the confusion created by subjective terms. "Cognition" sounds to the uninitiated like a conscious mental process. But the psychologists who have led the recent revival of interest in this central nervous activity (e.g., Terrace 1984) do not mean any such thing. For them, consciousness is quite another matter and Cartesian dualism is a "spectre" *not* being raised (Terrace 1984).

Another animal psychologist, Gallistel (1980), has presented a fresh synthesis of behavioral causation challengingly entitled *The Organization of Action*, which is based explicitly on Sherringtonian reflex physiology. Tactically perhaps, he employs a less "dated" expression than reflex action, namely, "sensorimotor coordination," but this is an exact synonym. To an outsider it looks something like a U-turn in the history of their subject when we find psychologists (zoologists generally are still lagging here) turning back to build anew upon the achievements of the pioneer behavioral physiologists: not only Sherrington, but also Pavlov (see Staddon 1983; Mackintosh 1983), von Holst (see Gallistel 1980), Weiss (see Gallistel 1980), and others. Their achievements were effectively set aside for many decades by psychological, ethological, and even some physiological students of animal behavior. Witness Roeder writing in 1962:

> Old standbys, such as the reflex and the morphological center, have become peripheral in their significance. (Roeder 1962, p 115)

On the other hand concurrent advances in neurophysiology quickly made nonsense of the supposed lack of endogenous activity and the related idea of a fixed, one-to-one relationship of input to output in reflex action. Suffice it to recall Roeder's own forceful statement:

> To consider means whereby endogenous factors are prevented from preempting the output of a neuron, so as to leave some control to sensory input, seems as important as it is to determine the basis of endogenous activity. (Roeder 1963, pp 119–120)

One could equally cite Dethier on central excitatory and inhibitory states, and others together with Sherrington quoted in Kennedy (1954, 1966).

Gallistel (1980) enlarges on the *hierarchical* organization of sensorimotor coordination, not as a top-downwards cascading system of motivating impulses such as Tinbergen's (1950) original one, but a "lattice hierarchy" of integrative levels (cf. Dawkins 1976). In particular Gallistel elaborates the point that similar coordinating principles—Staddon's "rules of interaction"—can be recognized in operation at many integrative levels, coordinating larger and larger units as one ascends the hierarchy (cf. Sherrington 1947, p 238; Kennedy 1958; Manning 1967):

> The problem of motor coordination becomes the problem of motivation as one ascends the action hierarchy. (Gallistel, 1980, p 287)

> The principles governing waxing and waning of potentiation at motivational levels in the hierarchy are much the same as those encountered at lower levels. (Gallistel, 1980, p 332)

If motivation is not something new that comes in only at higher levels, then it becomes an unnecessary concept just as it did for Dethier from his own experimental evidence.

The knee-jerk, the recoil from a hot stove, and a few other instant responses are cited routinely to illustrate the supposedly simple, push-button character of reflexes. But they are not typical. They do of course show relatively constant input-output relationships and may be relatively simple in having short first-order arcs through the spinal cord. But they also have collateral arcs up through the brain that can modulate and even inhibit them, as when someone acts bravely under the dentist's drill. They are what Sherrington called *strong* reflexes that override all their rivals, unlike *weak* reflexes such as grooming, which are easily overridden. A reflex can be innately strong, as with the hot stove; or it can become strong through social reinforcement, such as recoil from a cockroach; or it can come out strong simply because the stimulus is "supernormal," such as the sudden stretch of the patellar tendon that produces a knee-jerk. Strong reflexes are but one end of a long spectrum.

In sum, nonmotivated behavior has no physiological reality, so there is no dichotomy of behavioral mechanisms. The gross misapprehension of reflex action sustains a false dichotomy between it and "higher" modes of behavior.

Survival of the Unfit

How is it, then, that the dichotomy has lived on to this day? When the pioneer ethologists' observations became widely known in the 1950s (Lorenz 1950; Tinbergen 1951), it was at once obvious to everyone that they could not conceivably be accommodated by the push-button model of behavior. There was undeniably something more, and the Mechanist behaviorists' vocabulary had no words for this enigmatic something. In the result, psychological terms once discredited by the behaviorists but reintroduced by the ethologists rapidly acquired scientific legitimacy among students of animal behavior including some physiologists (e.g., Bullock 1965; see Kennedy 1954, 1966). Just because these concepts were quite separate from reflex action as it was conceived, they did not displace the push-button caricature of reflexes but simply supplemented it. In this way the caricature lived on, along with the dichotomy, unscathed.

The most telling evidence for internal "energy" accumulation, drive, and motivation came from the ethologists' observations described as *vacuum* and *displacement* activities, for these are activities that occur "spontaneously," in the absence of normal external stimulation. They are glaringly nonreflex by definition on the basis of the push-button model. Tragically, no one seems to have noticed at the time that this phenomenon had been observed more than half a century before by Sherrington (1947, p 208) in spinal reflexes. He called such occurrences "spontaneous reflexes"; and that is not as self-contradictory as it sounds since the activities in question were normally observed as responses to stimuli, exactly paralleling the ethologists' observations of whole animals.

Spontaneous reflexes seem to depend on the principle of reflex interaction called *postinhibitory rebound* or *successive induction* (Sherrington), or, at the behavioral level, *antagonistic induction*. A reflex action can sometimes occur without the usual stimulus provided it has been kept inhibited for long enough by a stimulus that elicits an antagonistic reflex. This principle in the temporal coordination of different actions has now been recognized as operating at the neuronal and whole-animal levels of coordination as well as at the spinal reflex level (Staddon 1983, pp 36–46; Kennedy 1958, 1985). That this prior work is still left aside shows how strongly entrenched the push-button caricature of reflexes still is.

Staddon (1983) perhaps overlooked spontaneous reflexes when he wrote (p 30–31) "To be sure, there is always a stimulus. . . ." No doubt he meant "a stimulus" in the behaviorist's conventional one-sided sense of an external and excitatory one, but the effective stimulus for a spontaneous reflex is *inhibitory*, and this has obscured the reflex nature of the "spontaneous" action. Although "inhibition is co-equally with excitation a nervous activity" (Sherrington, in Denny-Brown 1939), our natural tendency as observers is to think only of the excitation—what the animal is doing—that being how we think of our own behavior. The necessary concurrent inhibition of everything the animal is not doing is then easily forgotten: out of sight, out of mind (Kennedy 1958, 1967). Yet what the animal does next depends not only on what new stimulus the ongoing behavior brings in to the animal (this alone is the old "chain-reflex" hypothesis mentioned by Dethier and Stellar 1961, p 70), but also on after-effects of the ongoing behavior including the "priming" of rival behaviors during their inhibition by the ongoing one: antagonistic induction.

The Cost of Convenience

Dethier, almost alone among current writers, perceived that the harm done by mentalist descriptions of animal behavior was not after all negligible. The price paid for their convenience is altogether too high when proximate causes are in question. As we have seen (p 18) mentalist terms confuse ultimate with proximate causes. Their convenience is a pitfall trap. Referring as they do to ultimate rather than proximate causes, they are devoid of physiological meaning. Their convenience aside, their role is only to make ultimate causes sound like proximate ones. Labelling each behavioral sequence of an animal by its end-result implies that this "explains" it in the proximate sense (see Tinbergen's remark quoted below, p 27). The unfortunate result is that direct analysis of the actual inputs and outputs seems less than urgent. Important features may then be missed (examples in Kennedy 1972, 1985, 1986a,b).

For instance, when what appear to be "irrelevant," "displaced" activities are attributed to "thwarting" or "frustration," this concentrates our attention on the *absence* of an "expected" stimulus, because the absence of it is uppermost in our minds in a like situation. Other stimuli that are present and capable of eliciting quite different activities are out of mind. So too, therefore, is the possibility that one of them, being now disinhibited, will appear with the aid of postinhibitory rebound. This possibility was in fact raised by the original observation that "displaced" activities are abnormally intense. The observation has recently been confirmed quantitatively in rats, but was passed over when the original "overflowing energy" theory of displacement was replaced by disinhibition theories (see Kennedy 1985).

Mentalist terms as we have seen:

1. Sustain a gross misapprehension of real reflex action.
2. Thereby sustain a false dichotomy between reflex action and "higher" modes of behavior.

3. That dichotomy in its turn has deprived ethologists of the help they could have had from real reflex physiology.
4. Worst of all, the dichotomy has kept wide open what Roeder (1965, p 250) called "the gap between what nerve cells do and how animals behave."

Hopes were high in the 1950s that this great gap would be closed in the foreseeable future (Tinbergen 1950). But authors of present textbooks agree that there has been disappointingly little progress toward closing it (e.g., Hinde 1970, p 170; Halliday and Slater 1983, p 2). To do that required first a meeting of minds across the gap, but the mentalist language of ethology has obstructed this, being alien to physiologists (see below).

Physiologists and neuroethologists have made considerable progress on their side of the gap, working on a number of lower-level systems in the behavioral hierarchy (reviews by Guthrie 1980; Ewert et al. 1983; Huber and Markl 1983; Camhi 1984). But meanwhile discouraged ethologists have found it more profitable to abandon bridge-building, and go over to abstract modelling, ultimate causation, or human behavior. In those fields mentalist terms have a still freer rein; and this in turn helps to maintain similar language in proximate causal studies as well since we all take an interest in both proximate and ultimate causes. It is a vicious circle.

Subconsciousness

What Dethier said of psychologists is equally true of zoologists using psychological language habitually:

> Physiological psychologists . . . have not purged their subconsciousness of human subjective and philosophical underpinnings. (Dethier 1981, p 463)

"*Sub*consciousness" is the operative word. This is Dethier's second major contribution: the effects of these "underpinnings" are insidious because so often they are not recognized—for instance by Keeton (p 20).

As to bridging the "gap," it did not help when Sherrington himself (1947, new Foreword), late in life, took up an outright dualist stance and insisted that the mindless reflex could not account for instinctive urges and drives. Roeder was intrigued as a physiologist by the ethologists' discoveries and unlike Sherrington kept his feet on the physiological ground. He wrote (1962) of the frustration he felt for lack of some synthetic concepts with which to move upwards from his neuronal level of analysis toward complex behavior. He worried about where the needed concepts could come from. Evidently he was unable to make use of the ethologists' psychological concepts for his purpose: they posed no clear questions that a neurophysiologist could tackle. He could not turn to reflex theory as the alternative, having misapprehended and dismissed it as a matter of push-buttons (see also Kennedy 1967). A neuroethologist of the following generation is still complaining of the difficulties presented by the vague unphysiological concepts of ethology (Guthrie 1980, p 18); while an insect neurophysiologist of the following generation echoes Roeder on the lack of synthetic concepts (general principles) and now, sad to say, unlike Roeder he seems resigned instead of worried about it (Burrows 1984).

Tinbergen, on the other hand, brought out many years ago most of the anti-anthropomorphist arguments advanced here, except that he too misapprehended reflex physiology as "grotesque simplification" (1951, p 101), thereby denying himself the help it could give in behavior analysis. After all, Sherrington worked on the behavior of limbs, not of neurons. Like Dethier, but unlike Lorenz and their joint scientific descendants, Tinbergen clearly recognized that the mentalist language of ethology was a serious impediment. He warned repeatedly (1942, 1951, 1963) against what he called, prophetically, the "tenacious hold" of subjectivist teleology.

> A tendency to answer the [proximately] causal question by merely pointing to the goal, end, or purpose of behaviour . . . is seriously hampering the progress of ethology. (Tinbergen 1951, p 4)

> Our habit of giving names to systems characterized by an achievement, has made thinking along consistent analytical lines much more difficult than it would have been if we could have applied a more neutral terminology. (Tinbergen 1963, p 414)

"Thinking along consistent analytical lines" meant not confusing ultimate with proximate causes and accordingly "more neutral terminology" meant strictly objective language uncontaminated by subjective teleology. But, having said that, Tinbergen compromised and things went on as before. He did not proceed to recommend the use of such "neutral" terminology, finding it too "dry" and "noncommittal" (Tinbergen 1963). In this way he reopened the door to teleology. He could not offer any better system of objective terminology because he too had rejected the language of reflex integration. The only alternative then was to continue using functional, teleological language. To mitigate the effects of it that he deplored, all he could suggest (1963, p 414) was "to accept any frankly functional term, as long as this is done consciously"—in other words, trying not to forget for a moment that the term is metaphorical only.

That proviso was evidently asking too much and his warnings were consequently ineffective, since Dethier (1981) still finds "enormous resistance" to abandoning the concept of motivation. This is not surprising because even those contemporary writers (e.g., Krebs and Davies 1981) who do warn their readers that misunderstanding can come from confusing ultimate with proximate causes do so only in passing, not pressing the point home by citing recent examples. This is nowhere near enough to immunize students against anthropomorphism. Tinbergen's warnings might have had more effect had he pointed out the *subconscious* nature of the anthropomorphism loading such terms, which Dethier did. Few workers today deliberately impute motives to animals. What Dethier and I are saying is rather that the subconscious effect of habitually using mentalist terms even metaphorically, is sooner or later to mistake ultimate causes for proximate ones.

Liberation?

When Dethier deliberately translated psychological descriptions of "motivation" into a "dry" objective one (p 19) he was, in effect, demonstrating how to avoid the pitfall of subconscious anthropomorphism, the problem Tinbergen left unsolved. This should permit the resumption of bridge-building from the behavioral side of Roeder's great "gap," to meet eventually the physiologists' building from their side, and with more chance of success than Tinbergen (1950) had in his pioneering attempt. Dethier's rejection of animal motivation may make more impression than Tinbergen's because, unlike Tinbergen (1942), he started out content with motivation and hostile to the anti-anthropomorphists; then hunted experimentally as no one else has for the motivating Snark, failed to find it, and was eventually driven to the conclusion that such concepts positively hinder understanding and should be abandoned. This principled return to the now un-fashionable position of the anti-anthropomorphists who goaded him into the Snark-hunt in the first place, is what earns Vince Dethier the title of liberator.

He has freed the proximately causal analysis of animal behavior from the false dichotomy between motivated and nonmotivated, reflex behaviors, leaving us with a single, coherent system to deal with. It is a system that uses similar unit mechanisms throughout but generates emergent properties through its hierarchical organization. The problem as Dethier found is consistently *physiological*. There is another false dichotomy in the tradition that "physiology" refers to the working only of parts of the body, as if behavior were something other than physiology and not simply the highest integrative level of it (Kennedy 1967). It was therefore no accident that Pavlov (1928, p 213 ff.) preferred to call behavior "higher nervous activity," although this again was asking too much. The approach is conveyed equally by Sherrington's "reflex integration," Gallistel's "sensorimotor coordination," and von Holst's "behavioural physiology." Notwithstanding Keeton (1967, see p 20), and likewise Thorpe's (1954) belief that "to say 'all behaviour is reflex' is stultifying for research," it does in fact help us to think of all behavior as in principle reflex—in the full sense of the term.

Certainly, the liberation does face stubborn resistance; and subjective language is so vivid and convenient for us that it cannot be given up altogether. And yet, the tide might just be turning in favor of the liberation, given the new currents in psychology flowing much the same way. There is the return to build anew on the neglected achievements of Sherrington, Pavlov, and other pioneer behavioral physiologists (p 23). Psychologists have recently demonstrated astonishing "cognitive" ("map-making") ability in animals (e.g., Herrnstein 1985), but at the same time they have widened the cognitive gap between ourselves and even our nearest relative the chimpanzee in the matter of language (Terrace et al. 1979; Humphrey 1983; Terrace 1984, 1985), thus rendering the imputation of a conscious mind and motives to any animal less plausible than ever.

Although "cognition" is another mentalist term, sometimes explicitly so (e.g., several authors in Mellgren 1983), it has not been shown that animal "map"-making is conscious (p 22). Nor have we any grounds for believing that, to account for our conscious minds, we have to resort to Cartesian dualism and postulate a *separate class* of proximate cause, any more than we have to postulate one to account for "motivated" behavior in animals. We know the capacity for "map"-making and -using has been elaborated in the course of evolution (it was first recognized under names such as "locality imprinting" in insects and "latent learning" in rats: Tinbergen 1951; Thorpe 1963), and this process has evidently continued in man's evolution to produce an extraordinarily sophisticated capacity for it, layer upon layer.

There is of course one respect in which motivation and reflex integration do fall into quite separate classes in ourselves, and that is simply the kind of evidence we have about them: introspective for motivation, etc., as against exteroceptive for the physiological mechanisms. We have still virtually no idea of how to relate the two pictures that we see from these two profoundly different viewpoints. Zoologists will help to solve that age-old problem, too, by setting definite bounds to it, if they drop the idea of motivation, and the term, as applied to organisms other than ourselves. The problem of consciousness can hardly be tackled while it seems as boundless as it does now. Just possibly, and thanks to Dethier in particular, that liberating step may now be in sight, reviving hope of closing the gap between what nerve cells do and how animals behave.

Acknowledgments I am grateful for comments and criticisms to Drs. C.E.J. Kennedy, C.J. Kennedy, P.L. Miller, T.J. Roper, S.J. Simpson, J.E.R. Staddon, and H.S. Terrace, to members of the Animal Behaviour Research Group in the Department of Zoology, Oxford, and to Vince Dethier himself.

References

Baker RR (1982) Migration Paths Through Time and Space. Hodder & Stoughton, London

Bullock TH (1965) Mechanisms of integration. In: Bullock TH, Horridge GA (eds) Structure and Function in the Nervous Systems of Invertebrates I. WH Freeman, San Francisco, pp 253–351

Burrows M (1984) The search for principles of neuronal organization. J Exp Biol 112:1–4

Camhi JM (1984) Neuroethology: Nerve Cells and the Natural Behavior of Animals. Sinauer Associates, Sunderland, Massachusetts

Collier GH (1980) An ecological analysis of motivation. In: Toates FM, Halliday TR (eds) Analysis of Motivational Processes. Academic Press, London, pp 125–151

Dawkins R (1976) Hierarchical organization: a candidate principle for ethology. In: Bateson PPG, Hinde RA (eds) Growing Points in Ethology. Cambridge University Press, Cambridge, pp 7–54

Denny-Brown D (1939) Selected writings of Sir Charles Sherrington. Hamish Hamilton, London

Dethier VG (1964) Microscopic brains. Science 143:1138–1145

Dethier VG (1966) Insects and the concept of motivation. Nebraska Symposium on Motivation, 1966:105–136

Dethier VG (1976) The Hungry Fly. Harvard University Press, Cambridge, Massachusetts

Dethier VG (1981) Fly, rat, and man: the continuing quest for an understanding of behavior. Proc Am Philos Soc 125:460–466

Dethier VG (1982) The contribution of insects to the study of motivation. In: Morrison AR, Strick PL (eds) Changing Concepts of the Nervous System. Academic Press, New York, pp 445–455

Dethier VG, Stellar E (1961) Animal Behavior: Its Evolutionary and Neurological Basis. Prentice-Hall, Englewood Cliffs, New Jersey

Ewert JP, Capranica RR, Ingle DS (eds) (1983) Neuroethology. Plenum Press, New York
Gallistel CR (1980) The Organization of Action: A New Synthesis. Lawrence Erlbaum, Hillsdale, New Jersey
Griffin DR (1984) Animal Thinking. Harvard University Press, Cambridge, Massachusetts
Guthrie DM (1980) Neuroethology: An Introduction. Blackwell, Oxford
Halliday T (1983) Motivation. In: Halliday TR, Slater PJB (eds) Animal Behaviour 1: Causes and Effects. Blackwell, Oxford, pp 100–133
Halliday TR, Slater PJB (1983) Introduction. In: Animal Behaviour 1: Causes and Effects. Blackwell, Oxford, pp 1–9
Herrnstein RJ (1985) Riddles of natural categorization. Philos Trans R Soc Lond Ser B 308:129–143
Hinde RA (1970) Animal Behaviour: A Synthesis of Ethology and Comparative Psychology. McGraw-Hill, London
Huber F, Markl H (1983) Neurophysiology and Behavioral Physiology: Roots and Growing Points. Springer-Verlag, Berlin
Humphrey NK (1983) Consciousness Regained. Oxford University Press, Oxford
Keeton WT (1967) Biological Science. WW Norton, New York
Kennedy JS (1954) Is modern ethology objective? Br J Anim Behav 2:12–19
Kennedy JS (1958) The experimental analysis of aphid behaviour and its bearing on current theories of instinct. Proc Tenth Int Congr Entomol Montreal 1956, 2:397–404
Kennedy JS (1966) Some outstanding questions in insect behaviour. Symp R Entomol Soc London 3:97–112
Kennedy JS (1967) Behaviour as physiology. In: Beament JWL, Treherne JE (eds) Insects and Physiology. Oliver and Boyd, Edinburgh, pp 249–265
Kennedy JS (1972) The emergence of behaviour. J Aust Entomol Soc 11:168–176
Kennedy, JS (1985) Displacement activities and post-inhibitory rebound. Anim Behav 33:1375–1377
Kennedy JS (1986a) Migration, behavioral and ecological. In: Rankin MA (ed) Migration: Mechanisms and Adaptive Significance. Contributions in Marine Science, Austin, Texas, 27, Suppl pp 7–25
Kennedy JS (1986b) Some current issues in orientation to odour sources. In: Birch MC, Kennedy CEJ, Payne TL (eds) Mechanisms of Insect Olfaction. Oxford University Press, Oxford, pp 11–27
Krebs JR, Davies NB (1981) An Introduction to Behavioural Ecology. Blackwell, Oxford
Lorenz KZ (1950) The comparative method in studying innate behaviour patterns. Symp Soc Exp Biol 4:221–268
Mackintosh NJ (1983) Conditioning and Associative Learning. Clarendon Press, Oxford
Manning A (1967) An Introduction to Animal Behaviour. Edward Arnold, London
Mellgren RL (ed) (1983) Animal Cognition and Behavior. North-Holland Publishing Company, Amsterdam
Pavlov IP (1928) Lectures on Conditioned Reflexes 1 (Gant WH transl & ed). International Publishers, New York
Roeder KD (1962) Neural mechanisms of animal behavior. Am Zool 2: 105–115
Roeder KD (1963) Nerve Cells and Insect Behavior. Harvard University Press, Cambridge, Massachusetts
Roeder KD (1965) Epilogue. In: Treherne JE, Beament JWL (eds) Physiology of the Insect Nervous System. Academic Press, London, pp 247–252

Sherrington CS (1947) The Integrative Action of the Nervous System, with a New Fore-word by the Author and a Bibliography of His Writings. (Samson Wright, ed). Cambridge University Press, Cambridge

Staddon JER (1983) Adaptive Behavior and Learning. Cambridge University Press, Cambridge

Staddon JER (1986) The comparative psychology of operant behavior. In: Lowe CF, Richelle M, Blackman DE (eds) Behavior Analysis and Contemporary Psychology. Lawrence Erlbaum, Hillsdale, New Jersey, pp 83–94

Terrace HS (1984) Animal cognition. In: Roitblat HL, Bever TG, Terrace HS (eds) Animal Cognition. Lawrence Erlbaum, Hillsdale, New Jersey, pp 7–28

Terrace HS (1985) Animal cognition: thinking without language. Philos Trans R Soc Lond Ser B 308:113–128

Terrace HS, Petitto LA, Sanders RJ, Bever TG (1979). Can an ape create a sentence? Science 200:891–902

Thorpe WH (1954) Some concepts of ethology. Nature 174:101–105

Thorpe WH (1963) Learning and Instinct in Animals. Methuen, London

Tinbergen N (1942) An objectivistic study of the innate behaviour of animals. Bibl Biotheor 1:39–98

Tinbergen N (1950) The hierarchical organization of nervous mechanisms underlying instinctive behaviour. Symp Soc Exp Biol 4:305–312

Tinbergen N (1951) The Study of Instinct. Clarendon Press, Oxford

Tinbergen N (1963) On Aims and Methods of Ethology. Z Tierpsychol 20:410–433

Wigglesworth VB (1939) The Principles of Insect Physiology. Methuen, London

Wigglesworth VB (1966) Control of responsiveness. Symp R Entomol Soc Lond 3:110

Chapter 3

Plasticity in Control Systems for Insect Feeding Behavior

ALAN GELPERIN*

The insect nervous system has provided excellent material for cellular and biochemical studies of neural information processing and integration. In particular, the physiological study of chemoreception in insects is more advanced than in any other animal group, in no small measure due to the seminal stimulation provided by V.G. Dethier (1955, 1976). In addition, the comparative physiology of the neural control systems regulating feeding behavior also received great impetus from Dethier's work, particularly the viewpoint that insect–vertebrate comparisons could usefully illuminate general issues pursued by both groups (Dethier 1982; Moss and Dethier 1983; Dethier and Bowdan 1984). A very natural question arises as to the reality and extent of learned adjustments to the feeding control systems of insects. The remarkable learning skills of the honeybee provide a dramatic example of the sophisticated computational ability of some insect nervous systems and encourage the view that learning questions posed in ethologically relevant terms might reveal learning of a high order in other insect species, including flies. During the last 20 years there has been a veritable explosion of work with mammals on food-assessment learning mechanisms, using behavioral paradigms that are readily adapted to insects. Thus the stage is set to assess the learning abilities of flies and a variety of other insect species in ways that both pose the learning question in terms natural to the animal's Umwelt and facilitate comparisons with analogous results from the mammalian learning literature.

It is still true that the black blowfly, *Phormia regina*, provides the most completely described feeding control system outside the Mammalia; however, the locusts are coming on strong (Dethier 1976; Simpson and Bernays 1983). In spite of the wealth of information available on the reflex pathways and relevant stimuli comprising the *Phormia* feeding control system, evidence for reliable and robust learned modifications of fly feeding responses has been sparse

*Department of Molecular Biophysics, AT&T Bell Laboratories, Murray Hill, New Jersey 07974, U.S.A

until very recently (Fukushi 1983; McGuire 1984; Prokopy et al. 1985). The pioneering experiments of M. Nelson (1971) gave the first adequately controlled demonstration of associative and nonassociative modification of *Phormia* feeding in a classical conditioning paradigm that paired water or salt stimulation of the tarsal chemoreceptors with sugar stimulation of the labellar chemoreceptors. A major advance occurred with the demonstration that *Drosophila* could be associatively conditioned (Spatz et al. 1974; Quinn et al. 1974; Quinn and Dudai 1976; Dudai 1977). The neurogenetic analysis of *Drosophila* learning is proceeding apace (Tully 1984; Quinn and Greenspan 1984) and the behavioral assay for learning has very recently been perfected to the point where 95% of the flies show a conditioned avoidance response (Tully and Quinn 1985).

Since *Phormia* offers many advantages for physiological and pharmacological analysis of the neural control system for feeding (Long and Murdock 1983; Yetman and Pollack unpublished) and the powerful behavioral assay for learning in *Drosophila* is readily adapted to *Phormia*, T. Tully, T. McGuire, and I set out to test the learning ability of *Phormia* using the odor-shock associative learning paradigm that produced such impressive results with *Drosophila*. Before presenting a progress report on those ongoing studies, I will review briefly our current understanding of the neural control system for feeding in *Phormia* and the learning results with *Drosophila* that motivated our learning work with *Phormia*.

Feeding Control System

The regulation of caloric intake in *Phormia* results from the interplay of external and internal sensory inputs with a dynamic central synaptic processing circuit (Gelperin 1972; Bernays and Simpson 1982). Several populations of external chemoreceptors monitor the presence of attractive food odors or tastes. Several populations of internal proprioceptors monitor the state of distension of caloric reservoirs. The central synaptic circuit evaluates the internal and external sensory inputs dynamically depending not only on the current state of caloric energy reserves, but also on the phase of the circadian cycle (Hall 1980), the state of water balance (Barton Browne 1968), reproductive condition (Stoffolano 1973; Belzer 1978, 1979; Rachman 1980), and individual experience (Nelson 1971; McGuire, Tully and Gelperin unpublished).

The several sets of external chemoreceptors on the legs, mouthparts, and antennae of *Phormia* have been intensively investigated to determine the sensory code underlying taste or odor discrimination and the role of a particular receptor population in triggering or controlling food ingestion (Dethier and Hanson 1965; McCutchan 1969; Rachman 1979). The antennal olfactory receptors, for example, mediate taxic movements toward odor sources signaling food (Bowdan 1981) but do not influence ingestion once tarsal and labellar chemoreceptors have contacted the food (Dethier 1961). The tarsal and labellar receptors have been mapped anatomically (Wilczek 1967; McCutchan 1969) and the responses of the five sensory neurons serving each taste hair cataloged as mechanoreceptive (Wolbarsht and Dethier 1958), sugar-responsive (Hodgson 1957; Omand and Dethier 1969), water responsive (Evans and Mellon 1962a), and salt-responsive, the latter mediated by cells roughly separated as cation and anion receptors (Evans and Mellon 1962b; Gillary 1966; and see Hanson, this volume). Input from the tarsal receptors can arrest locomotion and elicit proboscis extension directed to the site of tarsal stimulation (Yetman and Pollack unpublished). Input from a single labellar sugar receptor can elicit patterned activity in a population of proboscis extension motoneurons (Getting 1971; Pollack 1977). Using the tip-recording method of Hodgson et al. (1955) one can record the chemosensory discharge while simultaneously monitoring muscle activity with electrodes implanted in identified muscles of the proboscis (Figure 3.1). Remarkably precise measurements on the minimal sensory input needed to activate the motor output have been made. These measurements are extremely valuable in guiding and constraining the models of central synaptic processing in the *Phormia* feeding control system (Getting and Steinhardt 1972; Pollack 1977).

If labellar input triggered by contact with food is adequate to overcome the current level of inhibitory feedback signaled by internal proprioceptors, ingestion commences. Food is pumped into the foregut by the cibarial pump in the proboscis. From the foregut food either enters the midgut or is diverted to a diverticulum of the foregut, the crop, for storage and later digestion. After ingestion has ceased, crop contractions deliver boli of food to the foregut from which they are delivered to the midgut by coordinated action of the foregut muscles and the proventricular (= cardiac) valve (Gelperin 1966a; Thomson 1975). Understanding these postingestive food movements is essential because the neural negative feedbacks to ingestion arise from sensors monitoring foregut distension (Gelperin 1966b; 1967) and crop volume (Gelperin 1971a). A pair of stretch receptors with axons in the recurrent nerve signal to the CNS the duration and extent of peristaltic expansions of the foregut lumen. Severing the input pathway from these foregut receptors to the CNS leads to the syndrome of overeating called hyperphagia, where intake may be limited only by the mechanical properties of the body wall in constraining crop expansion (Figure 3.2). Hyperphagic

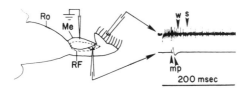

Figure 3.1. Procedure for simultaneously recording the chemosensory input and proboscis extension motor output. If a pipette containing a nonstimulating electrolyte and stimulating carbohydrate is applied to the tip of a largest labellar hair, activity in the water (w) and sugar (s) receptors will be recorded. A second electrode positioned on identified muscles involved in proboscis extension will record motor activity if the sensory input is sufficient to activate the central synaptic circuit for feeding. Individual motor units (m,p) can be resolved in the muscle recording. Ro, rostrum; RF, retractor of the furca; Me, mentum. (From Pollack GS. J Comp Physiol, 121: 115–134, 1977. Reproduced with permission.)

flies sometimes violate this constraint, hence these studies were the earliest work on the neural basis of bursting behavior (Dethier and Bodenstein 1958; Gelperin and Dethier 1967; Adams and Benson 1985). The stretch-sensitive neurons responsive to crop volume lie within the plexus of abdominal nerves emanating from the thoracicoabdominal ganglion to serve the abdominal segments (Bennettova-Rezabova 1972). The cells respond to stretch of the abdominal nerve branches consequent on crop filling (Gelperin 1971). Severing the neural pathway from these abdominal nerve stretch receptors to the CNS also induces hyperphagia. The conjoint operation that opens both negative feedback loops simultaneously produces a more severe and persistent hyperphagia than either operation alone. The interactions between the foregut and abdominal inhibitory inputs at different levels of food deprivation have been clarified by Bowdan and Dethier (1986), who provide evidence that the abdominal stretch receptors are important in ending each drink while the foregut stretch receptors provide a delayed and cumulative inhibitory feedback.

A general insight that emerged from these studies was the realization that a regulatory mechanism for caloric homeostasis could be constructed without using any sensors *directly* measuring caloric value. The external chemoreceptors

Figure 3.2. Hyperphagia in *Phormia*. The fly on the right was injected with pargyline whereas the fly on the left received a control injection. Both flies were then given access to 1.0 M sucrose. The hyperphagic fly ate more than twice as much as the control. Hyperphagia of similar magnitude is produced by section of the recurrent and median abdominal nerves. (From Long TF, Murdock LL. Proc Natl Acad Sci USA 80: 4159–4163, 1983. Reproduced with permission.)

are most effectively driven by stimuli that signal calorically useful foods whereas the internal proprioceptors signal only storage and movement of substances ingested at the command of the external receptors. The good correlation over evolutionary time of caloric density and chemoreceptor-coded acceptability is clearly sufficient. The few substances with high acceptability and low caloric value (e.g., fucose) are the exceptions that prove the rule. Although in the laboratory flies can be tricked into consuming nonnutritive fucose and starving to death in the presence of a nutritive glucose solution, their Umwelt contains essentially no fucose.

A summary diagram of the homeostatic mechanism for caloric energy flow in the blowfly is shown in Figure 3.3 (Gelperin, 1971b). The synaptic circuit in

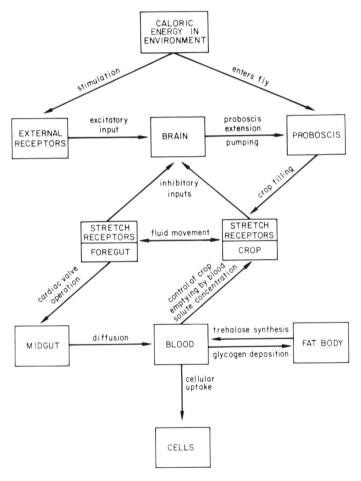

Figure 3.3. Homeostatic mechanism for the regulation of caloric energy flow in the blowfly. Note that the several populations of external chemoreceptors are indicated as a single population. The two populations of internal proprioceptors are shown associated with the structures that provide their activation. (Modified from Gelperin A. Annu Rev Entomol 16:365–378. © 1971 by Annual Reviews Inc. Reproduced with permission.)

the brain is shown receiving excitatory input from external chemoreceptors and inhibitory input from two populations of stretch receptors. The movement of food reserves from crop to midgut is controlled by the solute concentration of the blood, which is influenced by the equilibrium between cellular uptake of the insect blood sugar trehalose and its synthesis by cells of the fat body. In this way one can see how cellular energy utilization causes an increased probability of caloric energy ingestion (Gelperin 1972; Thomson and Holling 1977).

Olfactory Conditioning of *Drosophila*

The most recent experiments conditioning the fruitfly *Drosophila melanogaster* represent direct extensions of the initial instrumental learning paradigm of Quinn et al. (1974). The experiments involve pairing presentation of odor A with foot shock and pairing presentation of odor B with the absence of foot shock. A small group of flies is conditioned, simultaneously, and then tested by offering them a choice between moving into a tube containing odor A or into a tube containing odor B. A complete experiment consists of two reciprocal parts, the first of which involves pairing odor A with shock and the second of which involves pairing odor B with shock. Different groups of flies are used for the two reciprocal parts of a complete experiment. A learning index is calculated as the fraction of flies avoiding the shock-associated odor minus the fraction of flies avoiding the unshocked odor during the test cycle. The learning index for a complete experiment is taken as the average of the learning indices of the two reciprocal parts. The learning index is 0 if flies show no learning and 1.0 if all flies learn perfectly, avoiding only the shock-associated odor and showing no avoidance of the odor not associated with shock. The original implementation of the instrumental odor-shock training paradigm of Quinn et al. (1974) resulted in a learning index of 0.34 ± 0.02 for wild-type flies. The classical conditioning procedure of Tully (1984) results in a learning index of 0.91 ± 0.01.

Tully (1984) has presented an analysis of factors contributing to the dramatic improvement in learning performance evident in his experiments with *Drosophila*. A detailed description of the training and testing procedure, as implemented with *Phormia* using the apparatus shown in Figure 3.4A and B, will facilitate understanding the factors of general importance. The inner surface of the training tube is lined with an electrifiable grid and can be filled with odors of varying chemical species and concentration using an odor tube slipped over the end of the training tube. Air is drawn through the odor tube and into the training tube by a vacuum pump. An electronic stimulator delivers DC pulses of adjustable amplitude and duration to the grid floor of the training tube. Flies are trained and tested under dim red illumination to minimize phototactic bias.

To begin training, a group of flies is placed in the training tube and the stimulator set at the previously optimized parameters of 60 V and 1.25 sec pulse duration. An odor tube containing odor A (e.g., 3-octanol) is slipped over the end of the training tube and a vacuum line attached to the other end so that the training tube is filled with odor A. The flies are shocked every 5 sec for 60

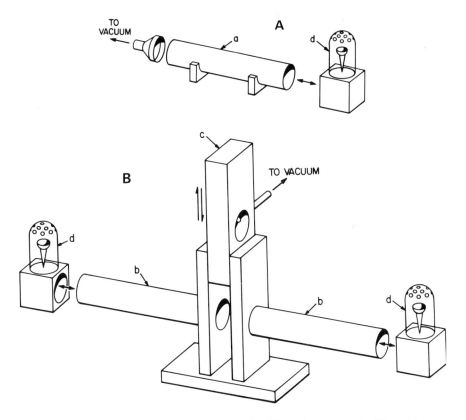

Figure 3.4. Classical conditioning apparatus for *Phormia regina*. A. The training apparatus consists of (a) a training tube, in which 95% of the inner surface is covered with an electrifiable grid, and (d) an odor tube containing an odor cup. B. The testing apparatus consists of (b) two collection tubes at a choice point for testing odor responses and (c) a sliding center compartment used to transfer trained flies to the choice point. A vacuum line is connected to a port on the center slider. Odor tubes (d) house odor cups containing the odorants used during training and testing. The odor cups slip onto the ends of either the training tube or the testing tubes.

sec. After odor A-shock pairing, the odor A tube is replaced with the tube containing odor B (e.g., 4-methylcyclohexanol) for 60 sec without shock. The entire training procedure produces minimal disturbance of the flies.

After one or more training cycles, the flies are moved into the sliding center compartment of the testing apparatus. While the freshly trained flies spend 90 sec in the sliding central compartment, the tubes containing odors A and B are slipped onto the ends of the test collection tubes. The test begins when the sliding central compartment is brought into register with the choice point, directly between the two test collection tubes, one connected to odor A and the other connected to odor B. Suction applied at the choice point draws air through both odor tubes, into the collection tubes and out through the center compartment. Flies are allowed to distribute themselves into the test collection tubes for 120 sec, at which point the sliding center compartment is displaced, trapping flies in the test collection tubes. The numbers of flies in each collection tube and in the central chamber are then counted. The reciprocal half of this experiment involves training a new group of naive flies using 4-methylcycloh-exanol as odor A and 3-octanol as odor B. The learning index is calculated as the fraction of flies avoiding the shock-associated odor minus the fraction of flies avoiding the unshocked odor, averaged for the two reciprocal parts of the experiment.

With shock pulse parameters held at 60 V DC and 1.25 sec duration, acquisition of the classically conditioned odor aversion response by *Drosophila* is a smooth function of the number of shock pulses delivered during training (Figure 3.5). Avoidance reached asymptotic levels with 10 shocks per training session, resulting in a learning index of 0.91 ± 0.01, which means that 95% of the flies avoided the shock-associated odor. Five additional training cycles did not improve performance (Tully 1984).

Even though the procedure of averaging the learning index derived from two reciprocal experiments, the first pairing odor A and shock and the second pairing odor B and shock, cancels out nonassociative effects, independent tests are needed to identify nonassociative factors that may be present before cancellation. Tully examined a sensitization control group exposed to odors without any shock, a pseudoconditioning group exposed to shock without any odor present, and an unpaired group presented with odor and shock separated by 30 sec. After these three nonassociative control procedures, one group of flies was tested for its avoidance of odor A versus air, and the other group was tested for its avoidance of odor B versus air. Each of the three nonassociative control groups was tested in these two ways, and each control procedure pro-

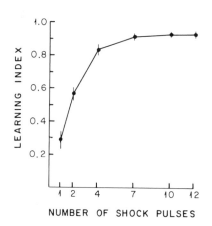

Figure 3.5. Learning by *Drosophila* as a function of the number of shock pulses delivered during training. Different groups of flies received varying numbers of 1.25 sec shock pulses during a single 60-sec training cycle. Mean learning indices ± standard errors are plotted for each group. Eight complete experiments were done for the first five groups, while 10 complete experiments provided the data for the 12-shocks group. (From Tully T. Behav Genet 14:527–557. Reproduced with permission.)

duced some decrement in odor avoidance responses relative to air. However, separate experiments using the same three nonassociative training procedures but testing for effects on odor avoidance choices between odor A and odor B presented *simultaneously* showed that learning indices were not affected by the presence of these nonassociative effects (Tully and Quinn 1985). These results, plus additional documentation of the retention, extinction, and reversal learning produced by the Tully procedure, make very clear the reproducible and robust nature of *Drosophila* associative learning (Tully 1984).

Olfactory Conditioning of *Phormia*

To investigate the ability of *Phormia* to learn odor-shock associations, the apparatus shown in Figure 3.4A and B was constructed, using a design similar to the *Drosophila* training machine but with larger dimensions to accommodate the larger body size of *Phormia* relative to *Drosophila*. For example the training and testing tubes were 1.75 in. in diameter and 9 in. long. The training tube was separate from the central slideway that conveyed the flies to the choice point during testing; hence in our experiments with *Phormia* the flies had to be transferred from the training tube to the central choice chamber after training. An electrifiable grid of dimensions suitable to the intertarsal dimensions of *Phormia* covered the interior of the training tube. The size of the odor tubes placed over the end of the training tube and over the ends of the testing tubes was also scaled up to accommodate the larger volume of the *Phormia* apparatus. Most of the experiments were done using the Lilly strain of *Phormia*.

A series of pilot experiments was performed to find an optimum pair of odors such that each odor would be reliably avoided when tested in a choice with moist room air and the pair of odors would give a 50:50 distribution of flies when the two odors were tested against each other. If both of the odors are reliably avoided when presented alone, the flies at the choice point are most likely to make a decision, i.e., move into one odor test tube or the other rather than remain in or return to the central choice chamber. If the degree of aversion to the two odors is well matched, then the decisions of the flies at the choice point should be a sensitive measure of any factor biasing their odor choices, e.g., learning. Several odors were screened, including 3-octanol, 4-methylcyclohexanol, amyl acetate, acetic acid, benzaldehyde (BAH), and methyl salicylate (MeSL). Of this group, BAH (10^{-3} M) and MeSL (10^{-4} M) gave the most encouraging results when tested alone and versus each other. On this basis they were selected for use in the initial series of conditioning experiments.

Flies were mass reared in cages with ad libitum access to solid sucrose, water, and pork liver. A group of 40–50 flies was aspirated from the stock cage just before training commenced. The entire group was transferred into the training tube within 5 min of selection. The training sequence consisted of 60 sec rest, 30 sec odor A, 60 sec rest, 30 sec odor B concurrent with 30 sec shock. This training cycle was repeated five times. Shock pulse amplitude was 60 V DC, pulse duration 0.5 sec, and repetition rate 0.2 sec^{-1}. With these parameters six shocks were delivered during each 30-sec odor-shock pairing. From the training tube the flies were transferred directly to the raised central chamber of the odor choice test apparatus and then rested for 1 min. After sliding the central chamber downward to convey the flies to the choice point, 90 sec were allowed for choices to be made, after which the central sliding chamber was displaced to trap flies in the two odor tubes and in the central chamber. The number of flies in each of the three compartments was counted. Again each complete experiment involved two reciprocal halves, with BAH associated with shock and MeSL unshocked in one half-experiment and MeSL shocked and BAH unshocked in the reciprocal half-experiment. Training and testing were conducted under uniform illumination from a 15-watt fluorescent bulb.

After 10 complete experiments involving over 1000 *Phormia*, a learning index of 0.32 ± 0.025 was obtained, with a range of values between 0.42 and 0.23 (McGuire, Tully, and Gelperin, unpublished). Two control groups testing for nonassociative effects were run with shock alone or odor alone during training. The learning indices of both these groups were not different from zero, indicating a negligible activation of nonassociative processes by the training paradigm. Five experiments were done with the third generation offspring of a wild female *Phormia* captured in Princeton, NJ. These wild Princeton flies yielded a learning index of 0.24, with a range from 0.35 to 0.12.

These results indicate that *Phormia* can show reproducible associative odor-shock conditioning. There are several reasons for thinking that the training and testing conditions are not yet optimized. The original Quinn et al. (1974) study of *Drosophila* learning involved screening 40 different odors to select two for intensive study. Such a wider screening of candidate training odors may well identify an odor pair whose use will lead to improvements of the *Phormia* learning index, perhaps by reducing the 10–12% of flies found in the central choice chamber at the end of the test. Several variations on the timing parameters in the training and testing protocol are needed, particularly testing for a beneficial effect of waiting 30–60 min between training and testing. We have reason to believe that the air flow in the central choice chamber of the *Phormia* apparatus is less than optimal. A new apparatus is planned with a smaller volume central chamber. A design that incorporated the training chamber into the central sliding section of the testing apparatus as in the original *Drosophila* training machine would also be desirable. There is ample reason to believe that with further evolution in experimental design the *Phormia* learning index may soon approximate that already demonstrated for *Drosophila*. This expectation is strengthened by the recent demonstration of two-trail color-food learning by the Australian sheep blowfly, *Lucilia cuprina* (Fukushi, 1985)

General Discussion

Twenty years ago the paucity of learning data on Dipterans led Dethier to speculate that flies may have lost the ability to learn in order to make more efficient use of their neural circuits for reflexive computations (Dethier 1966). Since that time the application of the highly developed methodologies of learning psychology to the search for modifications of feeding responses has revealed strong effects not only in flies, but also locusts (Jermy et al. 1982) and caterpillars (Dethier 1980). The learning abilities of insects have been so widely demonstrated that ecological theorists have begun to address the influence of learning on the stability of insect populations (Hassell and May 1974; Taylor 1974; Kacelnik and Krebs 1985).

With strengthened conviction that *Phormia* can display reliable, and ultimately robust, associative conditioning, great interest attaches to recent efforts to bring the modern tools of intracellular recording, stimulation, and cell staining to the interneurons comprising the CNS integrating circuits (Yetman and Pollack, unpublished; Brookhart et al. 1985). By applying a solution of cobalt lysine to the tips of labellar taste hairs for many hours, Yetman and Pollack (unpublished) and Edgecomb (1985) were able to trace the axons of the taste receptors to the subesophageal ganglion and describe the geometries of central branching of the receptor axons. Such precise mapping studies delimit the area of search for chemosensory integrating interneurons. The identification of octopaminergic synapses whose pharmacological manipulation dramatically and specifically affects feeding responses in *Phormia* (Long and Murdock 1983; Long et al. 1986; McGuire and Friedman 1985) opens the way to immunocytochemical mapping studies as another approach to localizing chemosensory interneurons. The neural microprocessor controlling *Phormia* feeding may yet yield its glowing secrets to illuminate the world of flies and men (Dethier 1981). If pursued with the creativity and breadth of insight evidenced by Vincent Dethier, the work will also help ameliorate the Beacon Hill syndrome characterizing the current state of communication between vertebrate and invertebrate neuroscientists, wherein "The Cabots speak only to the Lowells and the Lowells speak only to God" (Dethier 1982).

Another fundamental contribution to be gained from analysis of the neural processing of odor and taste inputs in the CNS of *Phormia* and other insects is the testing of new principles of neural computation (Hopfield 1982; Gelperin et al. 1985; Hopfield and Tank 1986). These new principles of neural computation provide a means of understanding the input-output functions of highly complex and interconnected neural circuits without access to the entire catalog of detailed local circuit interactions. Given the extreme pressures for miniaturization and efficiency exerted on the insect CNS over its long evolutionary development, one can reasonably expect to find neural microprocessing modules that provide both a rigorous testing ground for the new principles of neural computation and allow the testing of their generality.

Acknowledgments I thank K. Delaney, T. McGuire and T. Tulley for comments and J. Pollack, S. Yetman, R. Edgecomb, T. Long, and L. Murdock for prepublication access to their papers. Work at Princeton University was supported by NIMH Grant 39160.

References

Adams WB, Benson JA (1985) The generation and modulation of endogenous rhythmicity in the *Aplysia* bursting pacemaker neuron R15. Prog Biophys Mol Biol 46:1–49

Barton Browne L (1968) Effects of altering the composition and volume of the hae-
    molymph on water ingestion of the blowfly, *Lucilia cuprina*. J Insect Physiol 14:1603–
    1620
Belzer WR (1978) Recurrent nerve inhibition of protein feeding in the blowfly *Phormia
    regina*. Physiol Entomol 3:259–263
Belzer WR (1979) Abdominal stretch in the regulation of protein ingestion by the black
    blowfly *Phormia regina*. Physiol Entomol 4:7–13
Bennettova-Rezabova B (1972) The regulation of vitellogenesis by the central nervous
    system in the blowfly *Phormia regina*. Acta Entomol Bohemoslov 69:78–88
Bernays EA, Simpson SJ (1982) Control of food intake. Adv Insect Physiol 16:59–118
Bowdan E (1981) Activity of female blowflies *(Phormia regina)* in response to novel
    odours. Entomol Exp Appl 29:297–304
Bowdan E, Dethier VG (1986) Coordination of a dual inhibitory system regulating feeding
    behavior in the blowfly. J Comp Physiol 158A:713–722
Brookhart GL, Edgecomb RSA, Murdock LL (1985) Brain biogenic amines and blowfly
    feeding behavior. Soc Neurosci Abstr 11:368
Dethier VG (1955) The physiology and histology of contact chemoreceptors of the
    blowfly. Q Rev Biol 30:348–371
Dethier VG (1961) The role of olfaction in alcohol ingestion by the blowfly. J Insect
    Physiol 6:222–230
Dethier VG (1966) Insects and the concept of motivation. In: Levine D (ed) Nebraska
    Symposium on Motivation. University of Nebraska Press, Lincoln, pp 105–136
Dethier VG (1976) The Hungry Fly. Harvard University Press, Cambridge
Dethier VG (1980) Food aversion learning in two polyphagous caterpillars, *Diacrisia
    virginica* and *Estigmene congrua*. Physiol Entomol 5:321–325
Dethier VG (1981) Fly, rat and man: the continuing quest for an understanding of be-
    havior. Proc Am Philos Soc 125:460–466
Dethier VG (1982) The contributions of insects to the study of motivation. In: Morrison
    AR, Strick PL (eds) Changing Concepts of the Nervous System. Academic Press,
    New York, pp 445–455
Dethier VG, Bodenstein D (1958) Hunger in the blowfly. Z Tierpsychol 15:129–140
Dethier VG, Bowdan E (1984) Relations between differential threshold and sugar receptor
    mechanisms in the blowfly. Behav Neurosci 98:791–803
Dethier VG, Hanson FE (1965) Taste papillae of the blowfly. J Cell Comp Physiol 65:93–
    100
Dudai Y (1977) Properties of learning in *Drosophila melanogaster*. J Comp Physiol
    114:69–89
Edgecomb B (1985) Neural correlates and regulation of feeding behavior in the blowfly
    Phormia regina. Ph.D. Thesis, Purdue University
Evans DR, Mellon DeF (1962a) Electrophysiological studies of a water receptor asso-
    ciated with the taste sensilla of the blowfly. J Gen Physiol 45:487–500
Evans DR, Mellon DeF (1962b) Stimulation of a primary taste receptor by salt. J Gen
    Physiol 45:651–661
Fukushi T (1983) The role of learning on the finding of food in the searching behavior
    of the house fly, *Musca domestica*. Entomol Gen 8:241–250
Fukushi T (1985) Visual learning in walking blowflies, *Lucilia cuprina*. J Comp Physiol
    157A:771–778
Gelperin A (1966a) Control of crop emptying in the blowfly. J Insect Physiol 12:331–
    345
Gelperin A (1966b) Investigations of a foregut receptor essential to taste threshold reg-
    ulation in the blowfly. J Insect Physiol 12:829–841

Gelperin A (1967) Stretch receptors in the foregut of the blowfly. Science 157:208–210

Gelperin A (1971a) Abdominal sensory neurons providing negative feedback to the feeding behavior of the blowfly. Z Vgl Physiol 72:17–31

Gelperin A (1971b) Regulation of feeding. Annu Rev Entomol 16:365–378

Gelperin A (1972) Neural control systems underlying insect feeding behavior. Am Zool 12:489–496

Gelperin A, Dethier VG (1967) Long-term regulation of sugar intake by the blowfly. Physiol Zool 40:218–228

Gelperin A, Hopfield JJ, Tank DW (1985) The logic of *Limax* learning. In: Selverston AI (ed) Model Neural Networks and Behavior. Plenum Press, New York, pp 237–262

Getting PA (1971) The sensory control of motor output in fly proboscis extension. Z Vgl Physiol 74:103–120

Getting PA, Steinhardt RA (1972) The interaction of external and internal receptors on the feeding behavior of the blowfly, *Phormia regina*. J Insect Physiol 18:1673–1681

Gillary HL (1966) Stimulation of the salt receptor of the blowfly. III. The alkali halides. J Gen Physiol 50:359–368

Hall MJR (1980) Circadian rhythm of proboscis extension responsiveness in the blowfly: central control of threshold change. Physiol Entomol 5:223–233

Hassell MP, May RM (1974) Aggregation of predators and insect parasites and its effect on stability. J Anim Ecol 43:567–594

Hodgson ES (1957) Electrophysiological studies of arthropod chemoreception. II. Responses of labellar chemoreceptors of the blowfly to stimulation by carbohydrates. J Insect Physiol 1:240–247

Hodgson ES, Lettvin JY, Roeder KD (1955) Physiology of a primary chemoreceptor unit. Science 122:417–418

Hopfield JJ (1982) Neural networks and physical systems with emergent collective computational abilities. Proc Natl Acad Sci USA 79:2554–2558

Hopfield JJ, Tank DW (1986) Neural circuits and collective computation. Science 233:625–633

Jermy T, Bernays EA, Szentesi A (1982) The effect of repeated exposure to feeding deterrents on their acceptability to phytophagous insects. In: Visser JH, Minks AK (eds) Proceedings of the Fifth International Symposium on Insect-Plant Relationships. Pudoc, Wageningen, pp 25–32

Kacelnik A, Krebs JR (1985) Learning to exploit patchily distributed food. In: Silby RM, Smith RH (eds) Behavioral Ecology. Blackwell, Oxford, pp 189–205

Long TF, Murdock LL (1983) Stimulation of blowfly feeding behavior by octopaminergic drugs. Proc Natl Acad Sci USA 80:4159–4163

Long TF, Edgecomb RS, Murdock LL (1986) Effects of substituted phenylethylamines on blowfly feeding behavior. Comp Biochem Physiol 83C:201–209

McCutchan MC (1969) Response of tarsal chemoreceptive hairs of the blowfly, *Phormia regina*. J Insect Physiol 15:2058–2068

McGuire TR (1984) Learning in three species of Diptera: the blowfly *Phormia regina*, the fruitfly *Drosophila melanogaster* and the housefly *Musca domestica*. Behav Genet 14:479–526

McGuire TR, Friedman L (1985) Octopamine may mediate central excitatory state response in the blowfly. Soc Neurosci Abstr 11:367

Moss CF, Dethier VG (1983) Central nervous system regulation of finicky feeding by the blowfly. Behav Neurosci 97:541–548

Nelson MC (1971) Classical conditioning in the blowfly *(Phormia regina)*: Associative and excitatory factors. J Comp Physiol Psychol 77:353–368

Omand E, Dethier VG (1969) An electrophysiological analysis of the action of carbo-
    hydrates on the sugar receptors of the blowfly. Proc Natl Acad Sci USA 62:136–
    143
Pollack GS (1977) Labellar lobe spreading in the blowfly: regulation by taste and satiety.
    J Comp Physiol 121:115–134
Prokopy RJ, Papaj DR, Cooley SS, Kallet C (1985) On the nature of learning in oviposition
    site acceptance by apple maggot flies. Anim Behav 34:98–107
Quinn WG, Dudai Y (1976) Memory phases in *Drosophila*. Nature 262:576–577
Quinn WG, Greenspan RJ (1984) Learning and courtship in *Drosophila:* two stories with
    mutants. Annu Rev Neurosci 7:67–93
Quinn WG, Harris WA, Benzer S (1974) Conditioned behavior in *Drosophila melan-
    ogaster*. Proc Natl Acad Sci USA 71:708–712
Rachman NJ (1979) The sensitivity of the labellar sugar receptors of *Phormia regina* in
    relation to feeding. J Insect Physiol 25:733–740
Rachman NJ (1980) Physiology of feeding preference patterns of female black blowflies
    (Phormia regina). The role of carbohydrate reserves. J Comp Physiol 139:59–66
Simpson SJ, Bernays EA (1983) The regulation of feeding: locusts and blowflies are not
    so different from mammals. Appetite 4:313–346
Spatz H-Ch, Emanns A, Reichert H (1974) Associative learning of *Drosophila melan-
    ogaster*. Nature 248:356–361
Stoffolano JG (1973) Effect of age and diapause on the mean impulse frequency and
    failure to generate impulses in labellar chemoreceptor sensilla of *Phormia regina*.
    J Gerontol 28:35–39
Taylor RJ (1974) Role of learning in insect parasitism. Ecol Monogr 44:89–104
Thomson AJ (1975) Regulation of crop contraction in the blowfly, *Phormia regina*. Can
    J Zool 53:451–455
Thomson AJ, Holling CS (1977) A model of carbohydrate nutrition in the blowfly *Phormia
    regina*. Can Entomol 109:1181–1198
Tully T (1984) *Drosophila* learning: behavior and biochemistry. Behav Genet 14:527–
    557
Tully T, Quinn WG (1985) Classical conditioning and retention in normal and mutant
    Drosophila melanogaster. J Comp Physiol 157A:263–277
Wilczek M (1967) The distribution and neuroanatomy of the labellar sense organs of the
    blowfly, *Phormia regina*. J Morphol 122:175–202
Wolbarsht ML, Dethier VG (1958) Electrical activity in the chemoreceptors of the
    blowfly. Responses to chemical and mechanical stimulation. J Gen Physiol 42:393–
    412

Chapter 4
Vertebrate Taste Receptors

Lloyd M. Beidler*

Nutrition is of utmost importance to all animals. Most utilize specialized chemoreceptors to aid them in food selection and consumption. Since the same group of chemicals is necessary for cell growth, maintenance, and function in these animals, there may be similarities in the properties of their chemoreceptors. Vincent Dethier devoted much of his scientific career to the study of insect contact chemoreceptors, particularly those of the blowfly. How similar are the functional properties of these receptors to those of the much studied rodents and man?

Latency of Behavioral Response

One of the most interesting studies of Dethier (1968) revealed that not more than 100 msec of receptor stimulation is necessary for the initiation of a blowfly proboscis extension. By comparison Halpern and Tapper (1971) and Halpern and Marowitz (1973) showed that 84 msec stimulation of the rat's taste receptor population is enough for a decision as to whether the rat should make another lick at a taste stimulus. Thus, the needed duration of neural activity for a rat behavioral response is comparable to that of the blowfly.

Receptor Number and Behavioral Response

Dethier (1968) showed that complete proboscis extension can be initiated by stimulation of but one receptor cell of the blowfly. In fact, only a difference of three nerve impulses within 100 msec stimulation is sufficient to determine whether the fly accepts or rejects the solution.

The response to chemicals as well as the physiological function may differ

*Department of Biological Science, Florida State University, Tallahassee, Florida 32306, U.S.A.

with taste bud location in man and rodents. Those innervated by the laryngeal nerve may help regulate water intake (Shingai and Ikuno 1980). Those at the back of the tongue innervated by the glossopharyngeal nerve may help initiate the swallowing reflex. Taste buds of the soft palate and the tongue anterior may be more related to food selection. However, even those taste bud populations differ in their chemical response profiles. In the rat the number of taste buds on the soft palate innervated by the petrosal nerve is slightly greater than that of the anterior tongue nerve. However, the soft palate responds much better to sugars, whereas the anterior tongue excels in its salt response as shown by Nejad (1986).

The total number of rat taste buds that need to be stimulated to initiate a behavioral response is not known. However, if nerve transection eliminates over 90% of the rat taste bud population, the preference thresholds are changed only moderately (Pfaffmann 1952).

Response–Concentration Functions

The increase in neural activity with stimulus concentration is remarkably similar in the blowfly and rodents. It follows a sigmoidal relationship when plotted semilogarithmically with a fairly linear function with the logarithm over the range of 0.01–1.0 M sucrose, for example (see Smith et al. 1983). A hyperbolic function best describes the total response range.

Application (Beidler 1954) of the mass action law to taste receptor stimulation results in the hyperbolic relation:

$$R = \frac{CKR_s}{1 + CK} \tag{1}$$

where K is the binding constant and R_s is the magnitude of saturated response at high concentrations. Application of this equation to the mean response of 30 blowfly tarsal hairs published by Smith et al. (1983, Figure 2) results in a $K = 11$. This is of the same order of magnitude as that calculated by Beidler and Tonosaki (1985) for the hamster.

Accurate electrophysiological measurement of human taste neural activity is difficult (see Diamant and Zotterman 1969). However, if it is assumed that the sensory unit associated with a measured just noticeable difference (JND) in sensation to two different stimulus concentrations is directly proportional to the magnitude of neural activity ($S = kR$, $S_M = kR_s$), then Eq. 1 transforms to:

$$S = \frac{CKS_M}{1 + CK} \tag{2}$$

where S is the intensity of sensory taste response as measured by JND values and S_M is the maximum intensity of response. Rearrangement of Eq. 2 reveals a linear relationship between C/S and C:

$$\frac{C}{S} = \frac{C}{S_M} + \frac{1}{KS_M} \tag{3}$$

The relative magnitudes of increments of taste responses to supra-threshold concentrations may be obtained from the measure of successive JND values. Lemberger (1908) measured the consecutive JND values for saccharin and sucrose solutions. Assuming that each JND step represents an equal increment of intensity of sensory taste response, one may plot the intensity of taste response versus log concentration as shown in Figure 4.1.

Equation 2 is sufficient to describe Lemberger's data with sucrose and with sodium saccharinate as well as JND data obtained in our laboratory using NaCl solutions.Successive JND steps were determined by the method of Harris and Kalmus (1949).

The importance of the rather good fit is not only that the adsorption theory can quantitatively account for psychophysical data, but also that the value of K, the binding constant of the reaction, can be obtained from psychophysical measurements. This constant is of great importance to the biophysicist since it is related to the strength of binding of the taste substance to the receptor surface. The magnitude of this constant for a given taste solution is shown in Table 4.1.

The value of 1740 for the equilibrium constant of the saccharin reaction is several hundred times larger than that for most other taste reactions studied to date. This indicates a much tighter binding of saccharin to the taste cell and

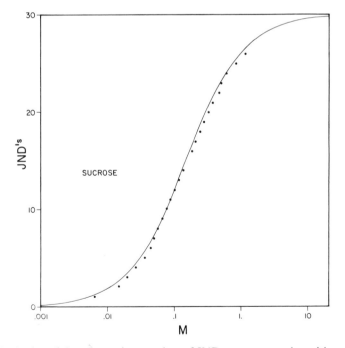

Figure 4.1. A plot of the successive number of JND steps versus logarithm of molarity of sucrose (M) as determined by Lemberger (1908). Solid line is calculated from Eq. 2 using $K = 6.6$ and $S_M = 30$ as obtained from data applied to Eq. 3. Solid circles represent experimental values.

Table 4.1. Threshold (C_T), binding constant (K) and maximum numbers of JNDs calculated from Eq. 2 as applied to published JND data

Taste solution	Theoretical C_T (mM)	Maximum number of JNDs		K
		Calculated	Experimental	
Sucrose				
Lemberger (1908) (human)	5.1	30	26[a]	6.6
Bujas (1937) (human)	6.8	16	13	9.2
Sodium chloride				
Beidler (unpublished 1955) (human)	2.7	29	26[b]	13.0
Bujas (1937) (human)	5.2	20	16	9.6
Saccharin				
Lemberger (1908) (human)	0.0115	50	41[c]	1740

[a]Higher concentrations became syrupy.
[b]Higher concentrations made patient nauseated.
[c]Higher concentrations stimulated other sense organs.

suggests the type of binding to be somewhat different from, for example, NaCl or sucrose.

It should be mentioned that the taste equation was originally applied to the steady levels of response when the taste solution was continually flowed over the tongue. Characteristics of the neural response other than that of the steady level may be of importance in the determination of psychophysical data. For

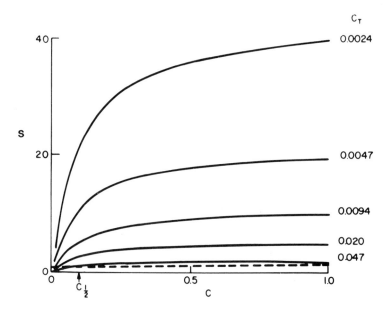

Figure 4.2. Equation 3 is plotted with intensity of sensory response (S) as ordinate and molar concentration (C) as abscissa. The value of K was assumed as 10 and S_M chosen as 40, 20, 10, 5, 2 for the highest to lowest curve, respectively. The threshold, C_t, was calculated from Eq. 6 for each S_M value.

example, a relatively high transient response to low concentrations of NaCl is observed in the neural activity, which declines within several seconds to a low steady level (see Beidler 1953, Figure 3). This transient response may be of greater importance than the steady level response when discrimination is required between two very weak taste solutions. Faull and Halpern (1972) made a detailed study of the features of the transient response. Also, in the derivation of Eqs. 1 and 2 it was assumed that taste intensity changed with concentration but that taste quality remained constant. However, it is well known that the sensation of taste of NaCl, for example, may change from sweet to salty at near-threshold concentrations. Both the transient response and the change in taste quality may alter the expected agreement between experimental data and the prediction of Eqs. 1 and 2 at very low concentrations.

The magnitude of K, the binding constant, in Eq. 2 is dependent only on the condition of the taste receptors and should not vary with changes of the central nervous system or the psychological state of the subject. If the intensity of the sensory taste response, S, varies with either of the latter two conditions, then the plot of S versus C also changes. Since K is constant, such changes result in a family of curves as illustrated in Figure 4.2. Although all the curves in Figure 4.2 appear quite different, the concentration necessary to elicit an intensity of response one-half of the maximum intensity is identical in all cases and has the value $C_{1/2} = 1/K$. We may use this knowledge to determine whether a change in the ability of a subject to discriminate taste solutions is a result of a variation in the binding of the molecules to the receptor or whether it is consequent on a change in the state of the higher nervous centers. The following conditions would result in a decrease in S_M with no change in K:

1. A decrease in the number of receptor cells stimulated by test solution.
2. A decrease in the number of sites per receptor cell.
3. A decrease in the ability of the receptor cell to respond to a given site when filled (efficacy).
4. An increase in the "noise level" of the central nervous system due to such factors as inattention or dominance of other sensory systems.

It is thought probable that conditions 2 and 3 would rarely appear and only under drastic physiological changes. Condition 1 can easily be controlled. Condition 4 is the most likely to occur and therefore controls the sensitivity of the taste sensory system as seen by changes in threshold, C_T.

Weber Fraction

The Weber relation states that $\triangle C/C$ = constant. This has been shown to describe most sensory data over a wide range of stimuli but may diverge at very low or very high stimulus values. A relation between $\triangle C$ and C may be mathematically derived from Eq. 3. If it is assumed that each JND unit represents an equal increment of intensity of sensory taste response, $\triangle S$, then:

$$\triangle S = \frac{KCS_M}{KC + 1} - \frac{KC_1 S_M}{KC_1 + 1}$$

Assign the arbitrary value of unity for the $\triangle S$ associated with a JND. Let $C - C_1 = \triangle C$ and solve for $\triangle C$:

$$\triangle C = - \left[\frac{1 + 2KC_1 + K^2C_1^2}{K(1 + KC_1 - S_M)} \right] \qquad (4)$$

or:

$$\frac{\triangle C}{C} = - \left[\frac{1 + 2KC_1 + K^2C_1^2}{KC(1 + KC_1 - S_M)} \right] \qquad (5)$$

The validity of using Eq. 5 to describe psychophysical data is indicated by the good fit of this theoretical curve to Lemberger's data as shown in Figure 4.3. Constancy of the Weber fraction is approached with moderate concentrations and digresses at very low and high concentrations. The divergence of the first few values at low concentrations is also dependent on the choice of concentration in the denominator of Weber's fraction. If this concentration is mid-point between the two values given by the $\triangle C$ used to determine the JND, then the first C value will always be equal to two since the higher concentration is of threshold value and the lower is zero concentration. On the other hand, if the higher of the two concentrations defining the first $\triangle C/C$ is used in the denominator of Weber's fraction, then the first $\triangle C/C$ value will always be unity. The value of the mid-concentration was used to calculate $\triangle C/C$ in Figure 4.3.

It should be noted that the value of S_M is based on the assignment of unity for a given JND. Therefore, S_M represents the total number of possible JND

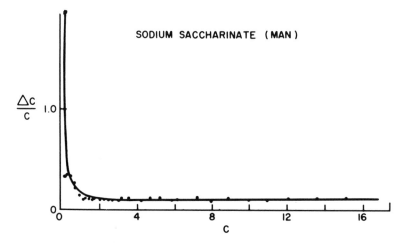

Figure 4.3. The solid line is the Weber fraction at different molar concentrations (x 10^4) calculated from the application of Eq. 5 using K and S_M of Table 4.1. The solid circles are actual values of $\triangle C/C$, as found experimentally by Lemberger (1908).

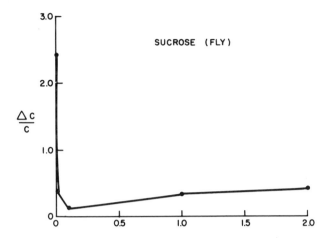

Figure 4.4. The Weber fraction, $\Delta C/C$, has been calculated from Dethier and Rhoades (1954) JND data for the response of the blowfly to sucrose and is plotted on the ordinate versus molar concentration on the abscissa.

steps for the given taste solution, on the chosen subject, and under the given physiological and psychological conditions. Table 4.1 shows calculated values for S_M and the total number of JND values found experimentally as well as the value of the equilibrium constant and the value of threshold calculated from Eq. 6 below.

If the Weber fraction is primarily dependent upon the intensity of neural response from the receptors as assumed in the derivation of Eq. 5 and does not involve further complications attributable to the central nervous system, then a similar Weber function should also be expected for lower forms of animals possessing simpler neural systems. Dethier and Rhoades (1954) measured the JND for the response of the common blowfly to sucrose at five different concentrations. Figure 4.4 shows that the Weber fraction for the fly varies with concentration in a manner very similar to the functions generated by Eq. 5.

Threshold Value and Its Relation to JND

As the concentration decreases, CK becomes much smaller than unity in Eq. 2 and S approaches unity, which is the condition for threshold. Therefore:

$$C_t = \frac{1}{KS_M} \tag{6}$$

Threshold is not a good indication of the ease with which the receptor can interact with the atoms or molecules of the taste solution. This is seen by the fact that K, which is a measure of the binding forces involved in the adsorption, is not directly related to the sensitivity of the receptors alone as determined by the threshold concentration, but sensitivity is also dependent on the maximum intensity of response, S_M.

Table 4.2. Threshold value C_T is not a good predictor of JNDs at higher concentrations

K	S_M	C_t	$(\triangle C)_{0.1}$	$(\triangle C)_{1.0}$
10	20	0.05	0.022	1.1
8	25	0.05	0.017	0.65
5	40	0.05	0.012	0.21
2	100	0.05	0.0073	0.045

Various values of K and S_M were chosen so as to obtain the same threshold from Eq. 6. The same values of K and S_M were used in Eq. 5 to obtain the incremental concentration for a JND at 0.1 and 1.0 M concentration.

The relation between the size of the increment in concentration for a JND, as shown by Eq. 4, and the magnitude of the concentration for threshold, as shown by Eq. 6, is not too apparent. Table 4.2 indicates that one may have different magnitudes of JND at supraliminal concentrations with the same threshold. The relationship between threshold and JND concentration is determined by the values of K and S_M, which may vary considerably if comparisons are made between two solutions of quite different taste qualities, for example.

Half-Concentrations

It has already been stated that the threshold value of concentration is of only limited interest. Is there any other concentration value of greater theoretical utility? The answer is yes, the half-concentration, $C_{1/2}$, that yields a magnitude of response exactly half of the maximum response. (The value of half-concentration, $C_{1/2}$, is based on the JND measure of response. It is not the value found by fractionation or similar psychophysical procedures.) It was previously stated that this half-concentration is numerically equal to the reciprocal of the binding constant. This can be shown by placing $C_{1/2} = 1/K$ in Eq. 2 and finding that this yields $S = S_M/2$.

If $C_{1/2} = 1/K$ is substituted in Eq. 5, then:

$$\frac{\triangle C}{C} = \frac{4}{(S_M - 2)} \tag{7}$$

Thus, the Weber fraction at this particular concentration value is a function of only the maximum response value, S_M. It can also be shown to be the minimal value by merely finding the slope of the $\triangle C/C$ versus C curve from Eq. 5 and setting it equal to zero. Thus:

$$\frac{d\left(\dfrac{\triangle C}{C}\right)}{dc} = \frac{-d\left[\dfrac{1 + 2KC + K^2C^2}{KC(1 + KC - S_M)}\right]}{dc} = 0$$

Solving for $C_{1/2}$

$$C_{1/2} = \frac{1}{K}$$

At half-concentration, $C_{1/2}$, the slope of Figure 4.1 can be shown to be dependent only on the value of S_M. Differentiating Eq. 2 with respect to C:

$$\frac{dS}{dc} = \frac{KS_M}{(1 + CK)^2}$$

Since $C_{1/2} = \dfrac{1}{K}$,

$$\left(\frac{dS}{dc}\right)_{C_{1/2}} = \frac{KS_M}{4} = \frac{S_M}{4C_{1/2}} \qquad (8)$$

Threshold is also related to half-concentration as shown by substituting $C_{1/2} = \dfrac{1}{K}$ into Eq. 6 and obtaining:

$$C_t = \frac{C_{1/2}}{S_M}$$

Intensity of Sweetness

It is often stated that saccharin is much sweeter than sucrose. A quantitative measure is utilized by comparison of their thresholds. However, we have already noted that threshold is inversely related to KS_M. Thus, two taste stimuli could have the same value of S_M and quite different values of K. In this case their thresholds would be greatly different but the taste stimuli would approach equisweetness at high concentrations. Schutz and Pilgrim (1957) demonstrated that the subjective intensity to humans of sweetness of saccharin and sucrose approach one another at high concentrations. Lemberger (1908), however, showed that the curve of consecutive JND values versus concentration almost saturated at 26 sucrose and 41 saccharin JND values. Does this indicate that JND values cannot be utilized for measurements of intensity of sweetness? Not necessarily, since there is no reason to believe that the size of a sucrose JND should be identical to that of saccharin. There is no neurophysiological or behavioral evidence to substantiate proof either way. Lemberger measured equal sweetness at various low concentrations of sucrose and saccharin and compared their aggregated JND values as shown in Table 4.3. Note that the saccharin JND values for all concentrations except the first are higher than that of sucrose by rather uniform factors of 1.5, 1.34, 1.52, and 1.57. At very high concentrations

Table 4.3. Aggregated JND values giving equal sweetness for sucrose and saccharin

Sucrose	Saccharin
4	4
6	9
10–11	14
12–13	19
14	22

From Lemberger F. Arch Gesamte Physiol 123:293–311, 1908.

the JND functions approach saturation with 26 JND values measured for sucrose and 41 for saccharin, again a factor of 1.57! This suggests the possibility of a constant of proportionality between sucrose and saccharin JND values for comparison of intensity of sweetness.

The above data suggest that man can discriminate at least 24 different sucrose concentrations between 0.01 and 1.0 M sucrose. Recently Smith et al. (1983) used information theory to calculate that the blowfly should theoretically be able to discriminate between 3.8 and 5.8 levels of sucrose over the same concentration range. Behavioral experiments using volumes of sucrose consumed at different concentrations as a measure of discriminability revealed that the blowfly can actually discriminate only about four levels of sucrose concentration as compared to 24 by man.

Equal Sweetness

It has been shown that JND magnitudes of sucrose and saccharin are not equal. However, neither the quality of sweetness nor the values of binding constant (6.6 versus 1740) are similar. In contrast, the binding constants of many monosaccharides are of the same order of magnitude. Can the JND values of these simple sugars be equated? Unfortunately, no measurements of consecutive JND values for these sugars are available in the literature.

Dahlberg and Penczek (1941) measured equal sweetness of glucose as compared to sucrose for nine different concentrations. This allows construction of a JND glucose curve as compared to Lemberger's (1908) JND sucrose curve. The values resulting from application of Eq. 2 are $S_M = 33$ and $K = 2.2$. If glucose and sucrose molecules compete for the same receptor sites, then the magnitude of response for the mixture is given by (see Beidler 1961):

$$S_t = \frac{(CKS_M)_s + (CKS_M)_g}{1 + (CK)_s + (CK)_g}$$

Both Dahlberg and Penczek (1941) and Cameron (1944) measured the concentrations of sucrose equal in sweetness to a series of sucrose–glucose mixtures. Utilizing the above equation with the previously found values for K and S_M for both sugars, the equivalent sucrose values can be calculated. The excellent agreement between observed and calculated values is shown in Table 4.4. These results suggest that the magnitude of JND values for glucose and sucrose may be nearly equal, in contrast to those of saccharin and sucrose.

An alternative model for taste intensity discrimination was proposed by Maes (1984). It is based upon signal-to-noise ratio. The standard deviation of the response is assumed to be a linear function of the magnitude of response. This is then incorporated into Eq. 1. The result is an almost hyperbolic relationship between response and concentration over a wide range of parameters. Since both Eq. 1 and the signal-detection derived function can support experimentally obtained behavioral data, Maes concludes that the good agreement with Eq. 1 is fortuitous, and similar agreement with the signal-detection derived function

Table 4.4. Concentrations of sucrose (%) that are equal in sweetness to various mixtures of glucose and sucrose

Mixture (%)	Sucrose of equal sweetness		Percent difference
	Observed	Calculated	
Dahlberg and Penczek (1941)			
10 S + 5.3 G	15.0 S	14.5 S	3.3
16.7 S + 8.3 G	25.0 S	25.0 S	0
26.7 S + 13.0 G	40.0 S	41.5 S	3.8
Cameron (1944)			
10 S + 5.5 G	15.0 S	15.0 S	0
5 S + 11.8 G	15.0 S	15.7 S	4.7
10 S + 10.15 G	20.0 S	19.3 S	3.2

is due to its sound theoretical basis! Unfortunately, he did not plot his function to indicate over what range of concentrations it can predict JND values as well as Eq. 1.

Summary

The chemoreceptor functions of man, rat, and fly are quite similar for many stimuli. The range of concentration over which they operate, as well as their concentration–response functions, are quite comparable. Thus, it is no surprise that the strengths of binding of the stimulus to receptor site are also similar.

Experiments indicate that both fly and rat can make a decision concerning preference within a 100 msec or less duration of stimulus contact.

All three species generate a similar Weber function. However, man is much better able to discriminate different levels of stimulus concentrations.

The utility of human JND measurements is much greater than previously thought. There appears to be a direct correspondence between JND and neural activity as a function of concentration. This provides a means of obtaining useful physicochemical information from behavioral data. In the future it may be possible to relate JND data to a variety of other psychophysical measures.

The half-concentration, $C_{1/2}$, defined as the concentration of taste stimulus that yields a magnitude of response one-half that of the maximum response, S_M, and S_M itself are both intimately related to many important psychophysical relationships. This is emphasized by summarizing:

$$\text{Threshold concentration, } C_I = \frac{C_{1/2}}{S_M}$$

$$\text{Binding constant, } K = \frac{1}{C_{1/2}}$$

$$\text{Weber fraction at } C_{1/2}, \frac{\triangle C}{C} = \frac{4}{(S_M - 2)}$$

The importance of chemoreceptor location in mammals is still not thoroughly understood. Evidence indicates that both response profile and physiological function may differ with location.

References

Beidler LM (1953) Properties of chemoreceptors of the tongue of rat. J Neurophysiol 16:595–607

Beidler LM (1954) A theory of taste stimulation. J Gen Physiol 38:133–139

Beidler LM (1961) Taste receptor stimulation. Prog Biophys Biophys Chem 12:109–151

Beidler LM, Tonosaki K (1985) Multiple sweet receptor sites and taste theory. Pfaff D (ed) Taste, Olfaction, and the Central Nervous System. Rockefeller University Press, New York, pp 47–64

Bujas Z (1937) La measure de la sensibilite differentielle dans le domaine gustatif. Acta Inst Psychol Univ Zagreb 2:1–19

Cameron AT (1944) The relative sweetness of certain sugars, and glycerol. Can J Res Sect E 22:45–63

Dahlberg AC, Penczek ES (1941) The relative sweetness of sugars as affected by concentration. N Y State Agric Exp Stn Geneva Tech Bull 258, 12 pp

Dethier VG (1968) Chemosensory input and taste discrimination in the blowfly. Science 161:389–391

Dethier VG, Rhoades M (1954) Sugar preference-aversion functions for the blowfly. J Exp Zool 126:179–203

Diamant H, Zotterman Y (1969) A comparative study on the neural and psychophysical response to taste stimuli. In: Pfaffmann C (ed) Olfaction and Taste III. Rockefeller University Press, New York, pp 428–435

Faull J, Halpern B (1972) Taste stimuli: time course of peripheral nerve response and theoretical models. Science 178:73–75

Halpern B, Marowitz L (1973) Taste responses to lick-duration stimuli. Brain Res 57:473–478

Halpern B, Tapper D (1971) Taste stimuli quality coding time. Science 171:1256–1258

Harris H, Kalmus H (1949) The measurement of taste sensitivity to phenylthiourea (PTC). Ann Eugen 15:24–31

Lemberger F (1908) Psychophysiche Untersuchungen über den Geschmack von Zucker und Saccharin. Arch Gesamte Physiol 123:293–311

Maes F (1984) A neural coding model for sensory intensity discrimination, to be applied to gustation. J Comp Physiol A 155:263–270

Nejad M (1986) The neural activities of the greater superficial petrosal nerve of the rat in response to chemical stimulation of the palate. Submitted to Chem Senses

Pfaffmann C (1952) Taste preference and aversion following lingual denervation. J Comp Physiol Psychol 45:393–400

Schutz T, Pilgrim F (1957) Sweetness of various compounds and its measurement. Food Res 22:1–8

Shingai T, Ikuno H (1980) Roles of laryngeal water fibers in water intake and urine flow in rats. J Physiol Soc Jpn 42:330

Smith D, Bowdan E, Dethier VG (1983) Information transmission in tarsal sugar receptors of the blowfly. Chem Senses 8:81–101

Chapter 5
Volta and Taste Psychophysiology

CARL PFAFFMANN*

As a Professor Emeritus, all of 3 years standing (having retired in 1983), I welcome our friend and colleague, Vince Dethier, to our ranks on reaching that blissful state of Emeritus. Like many of you, I have enjoyed his friendship and valued his scientific partnership over the years in the study of animal behavior and the chemical senses. My first efforts in that domain began as a graduate student in psychology at Brown University in 1933–1935, and then in England, ultimately at Cambridge, where I completed my Ph.D. dissertation on the electrophysiology of taste in the cat. In that physiology laboratory, as in many others, the cat was a presumed prototypic mammal. As it turned out, it is not a prototypic taster, for it seems not to have developed or retained sensitivity to sugars and other sweeteners that characterizes so many other species of mammals and invertebrates, especially, of course, the fly. I first became acquainted with the fly as a taster by Vince Dethier's work and in fact met him while he was still at the Hopkins where he began to "know a fly." Other longtimers there included Curt Richter, who by then "knew the rat," in particular its specific hungers and self-selection behavior, and Eliot Stellar at the Psychology Department. The Hopkins was a special place at that time in behavioral biology and psychobiology, especially with regard to ingestive behavior and its sensory determinants.

But I also remember Vince Dethier in other contexts, in his natural habitat at Blue Hill, Maine, or as a skier on various occasions, particularly his iconoclastic technique of making turns with his weight, not on the lower downhill ski, but on the upper ski. Many, if not most of us, were taught and continued to turn with our weight on the downhill ski. I recall a winter "Brain Research Conference" at Steamboat Springs, where we skied together and compared the relative grace and accuracy of our respective ski turns. I won't say who was the better, but I must add that I didn't even try to shift to the Dethier

*The Rockefeller University, 1230 York Avenue, New York, New York 10021, U.S.A.

method. To return to the present, I am glad to join in this celebration in his honor.

When I retired for the first time in 1978, as an academic Vice President at Rockefeller, I continued on as a Professor for another 5 years. I again had more time for research myself, "at the bench," instead of simply "laying on of hands" with graduate students, postdocs, and research fellows. One topic that at the time seemed still poorly understood and on which no one else seemed to be working or to be interested in was the phenomenon of "Voltaic taste," better known as electric taste.

Up until the beginnings of the electronic age, most functional studies of the senses were carried out by methods of psychophysics, that is the precise physical and chemical control of stimulation to a particular sense organ, the eye, the ear, skin, taste, olfaction, etc., coupled with the response of the human observer (detection and differential thresholds, quality identification, scaling of magnitude of sensation) or the behavioral responses of animals to controlled stimulation (proboscis extension, ingestion, self-selection). Except for anatomical studies, these might be considered indirect methods. Indeed, my introduction to the chemical senses was via a psychophysical study (Pfaffmann 1935) on a then new "method of single stimuli" for determining taste difference thresholds in humans. But I had read about the new developments in recording directly from the sensory nerves with awe and hope. Indeed I had the good fortune of a scholarship award that permitted me to go to Cambridge University and actually work in Lord Adrian's laboratory. His "basis of sensation" and many other important contributions are well known to this audience.

Volta in 1792 reported that dissimilar metals touching each other and the tongue surface gave rise to taste sensations. The anode gave rise to a sour taste, whereas the cathode, a sharp alkaline, possibly slightly bitter, sensation. Sulzer, in 1752, had in fact made a similar observation, but his explanation was in terms of a vibratory process occurring in the metals touching the tongue. Volta correctly attributed the effect to the electric fluid flowing through the tongue from one metal to the other. Terminating the anodal current, "anode break," led to a disappearance of the sourness, "cathodal break" caused a very brief period of sourness plus some sweetness.

The front of the tongue in humans appears to be more sensitive to the anodal stimulus than does the posterior tongue. One important classic observation by Oehrwall in 1891 showed a correspondence between the punctate chemical sensitivity of fungiform papillae when stimulated individually by chemicals and/or electric taste. Papillae sensitive to acid solutions reacted with a sour taste to the anode; papillae sensitive to bitter and sweet solutions could be activated by the onset of cathodal current. Thus a correlation between chemical sensitivity and electrical stimulation depending on polarity seemed established. I observed a similar concordance between chemical and electrical sensitivity to the anode in my recordings from the taste afferents in the cat (Pfaffmann 1941). A unit from the front of the tongue of the cat that responded to acetic acid showed a very similar response to anodal stimulation, except that the response had a shorter latency.

Electric taste employed as "electrogustometry" has been used as a clinical test with patients suffering abnormalities of taste (e.g., Krarup 1958; Frank et al. 1986). Neuropathy or accidental destruction of one of the taste nerves to the tongue shows up clearly in elevated or distorted electric taste thresholds. Krarup (1958), Fons (1976), and Hughes (1969) reported that electrical sensitivity changes with age, the older subjects having higher thresholds than the younger patients, but these changes are generally greater than those reported for chemical stimuli (Weiffenbach et al. 1982; Moore et al. 1982).

We did little further work with electric taste until Professor Bujas came to our lab as a National Academy of Sciences-Yugoslav Academy Exchange Professor 20 years ago. He, Marion Frank, and I then examined "voltaic taste" in the rat with electrophysiological recordings (1979). Figure 5.1 shows summator records of the electrophysiological discharges in the rat chorda tympani to acid and to anodal current applied to the anterior tongue via a weak subthreshold sodium chloride solution in the nonpolarizable stimulating electrode (Bujas, Frank, and Pfaffmann, unpublished). We noted a number of parallels between human psychophysical results and the chorda tympani responses in the rat. Anodal currents had the lowest threshold as they do for man, "cathode-off" slightly higher but "cathode-on" significantly higher. Units responsive to acid or to salt responded both to the anode-on and to the cathode-off. The number of salt- and acid-sensitive single units was greatly in excess of those responsive to sugar in our experiments but we had one or two sugar units that did respond

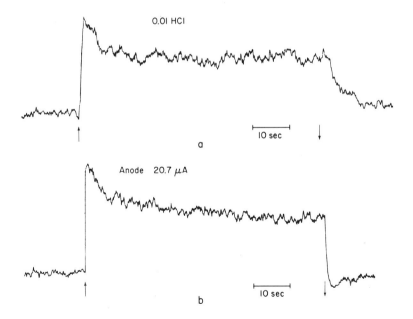

Figure 5.1. Summator responses of the whole chorda tympani of rat (a) to acid and (b) to anodal polarization. (After Bujas Z, Frank M, Pfaffmann C. Sensory Proc 3:353–365, 1979.)

to the "cathode-on." Acids and anodal currents give quantitatively equivalent responses.

Almost as soon as we began our studies of electric taste other investigators also showed a reawakened interest. Of course, Professor Bujas had continued psychophysical studies including electric taste at Zagreb (Bujas 1971) and there had been unpublished electrophysiological studies by Nejad over 20 years ago in Beidler's lab. Smith and Bealer (1975) had studied the rise time parameters of taste stimulation of anodal currents in which rate of rise can be readily controlled electronically. They showed that the time parameter was important in determining magnitude of the initial discharge. But a number of papers on electric taste appeared in the 1980 Japanese Taste and Smell Symposium. Kashiwayanagi et al. (1981) published "Similar effects of various modifications of gustatory receptors on neural responses to chemical and electrical stimulation in the frog." Ninomiya and Funakoshi (1980, 1981a, b) studied electrical stimulation in the rat chorda tympani single-units and the effect of various ions. Scott Herness (1985) completed a Ph.D. thesis on the biophysics of electric taste in Dr. Beidler's lab and Dr. Herness is now with me. So there is a renewed general interest in the subject, not simply because it is a historically exotic phenomenon, but because it does provide a more analytical way of studying the taste transduction process, about which there is still a considerable gap in our information.

Over the years two different hypotheses of the mechanism of electric taste have been proposed. One view assumes a direct effect of current on nerve fibers and taste cells. The other attributes stimulation to a chemical or electrochemical process in which the products of electrolysis or iontophoresis of saliva or extra- and intracellular fluids stimulate taste reception chemically. In our current experiments we have used a flow chamber for the tongue and silver chloride-coated wire to polarize the tongue surface (Figure 5.2). The indifferent electrode is also a nonpolarizable silver chloride wire making contact with the body tissues via an incision under the chin. Since we were interested not simply in sensitivity to the cations and anions of the common salts but in the sugar

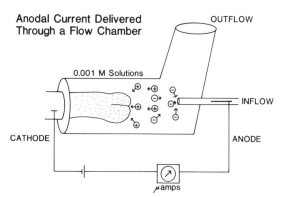

Figure 5.2. Schematic of tongue polarization chamber for anodal stimulation. (Courtesy of Dr. Scott Herness.)

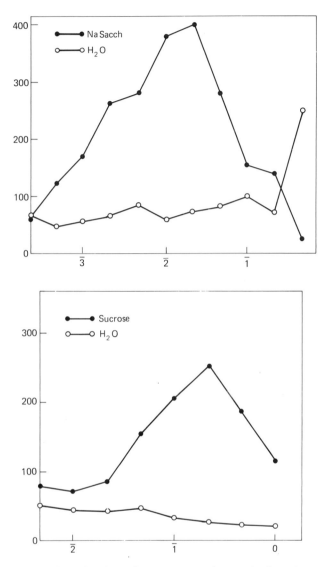

Figure 5.3. Hamsters' two bottle preference curves for saccharin and sucrose. Ordinate is total intake in milliliters for a group of nine males. Abscissa, log molar concentration. (After Carpenter JA. J Comp Physiol Psychol 49:139–144, 1956.)

sensitivity we employed the hamster instead of the rat. One of my students, Dr. Anthony Carpenter (1956), studied species differences in taste preferences among common laboratory animals. Figure 5.3 shows his behavioral results for the hamster, which clearly shows a strong preference, not only for sucrose, but also for sodium saccharin (so-called soluble saccharin or crystallose). *O*-Sulfobenzoic acid imide (insoluble saccharin) is much less soluble than its so-

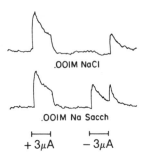

Figure 5.4. Summator records of hamster chorda tympani nerve responses to anodal and cathodal 5-sec pulses. Upper trace: Current was delivered via a 0.001 M NaCl adapting solution: Lower trace: Current was delivered via a 0.001 M Na saccharin solution. Note the slight decrement in response to cathode via NaCl and excitation via Na saccharin. Both traces also show a response to "anode-on" and a transient to "cathode-off." (Modified after Pfaffmann C. In: Pfaff DW, ed. Taste, Olfaction, and the Central Nervous System. Rockefeller University Press, NY, 1985.)

dium salt. I make a point of showing the behavioral reactivity of this species to both saccharin and sugar because there is considerable species variation so that a number of synthetic sweeteners do not taste sweet to nonhuman subjects. Electrophysiological studies had shown that the hamster and gerbil are much more reactive to sugar solutions than is the rat chorda tympani. The demonstration of a neural discharge to sweet stimulus, however, does not guarantee that that neural response is sweet to the recipient animal. Indeed, one can even extend the repertoire of behavioral methods to include the conditioned taste

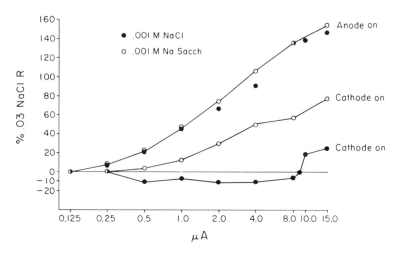

Figure 5.5. Magnitude of initial summator deflections to anode-on and cathode-on via 0.001 M solutions of NaCl and Na saccharin relative to the response to 0.03 M NaCl solution. (Reproduced with permission from Pfaffmann C, Pritchard T. In: van der Starre H, ed. Olfaction and Taste VII. IRL Press, London, 1980.)

aversion as we did (Nowlis et al. 1980). Hamsters conditioned to avoid sucrose also avoided fructose, glucose, *sodium saccharin,* and certain other nonsugar stimuli such as D-phenylalanine, all of which taste sweet to humans. The conditioned hamster did not avoid salts, acids, quinine solutions or cyclamate, or aspartame. The two latter synthetic sweeteners are not responded to as sweet by hamsters.

In any case, with this evidence we could safely proceed on the assumption that the electrolyte sodium saccharin was sweet, in order to compare the effect of using sodium saccharin versus sodium chloride as a bridging electrolyte for electric taste. The cation of sodium, either in sodium chloride or in sodium saccharinate, could be moved by iontophoresis to the tongue surface by anodal polarization. We can also compare the effect of iontophoresing the chloride anion versus the saccharin anion by cathodal polarization. Figure 5.4 shows in the whole nerve summated response the stimulatory effect of the anode at equal current for both solutions with the same cation, Na^+. The cathode, on the other hand, produces a mild inhibitory effect, that is, a reduction in resting activity for the chloride, compared with a stimulatory burst of activity in the case of the cathodal current with saccharin anion. These current levels are all far below those that stimulate the trigeminal touch, temperature, or pain receptors. In fact, control experiments by Dr. Tom Pritchard showed that single units of the trigeminal lingual afferents were not activated until much higher currents were utilized. Dr. Scott Herness' selective lingual or chorda desensitization leads to the same conclusions. Figure 5.5 shows the relationships between current intensity and magnitude of response quantitatively. The anodal response is essentially the same for the sodium chloride and the sodium saccharin. The cathodal responses are very different, however. Saccharin anion produces a positive response at low current values where the NaCl cathodal current results in inhibition.

If the stimulating effect of the electric polarization occurs by way of the cations and anions ónto the surface of the receptor, these cations and anions should show the same differences in stimulating capacity with electrical as with chemical stimulation. Species vary as to the order of efficacy of cations as taste stimuli. For the rat, hamster, and other rodents, sodium is more effective than potassium as a chemical stimulus. Similarly, for equal current values delivered via these two solutions, 0.001 *M* NaCl versus 0.001 *M* KCl, the electrophysiological responses to sodium is greater than to potassium. There is thus a parallel between the anodal electrical and cationic chemical stimulation. Since we made these observations, Dr. Herness, in his dissertation at Florida State University (Herness 1985), made a much more systematic study of the effect of various cations in relation to their ionic mobility and has shown a correspondence between anodal electrical stimulation and cationic efficacy as a taste stimulus for the rat chorda tympani.

The chorda tympani response to anodal polarization seems largely the result of stimulating salt and/or electrolyte cation-sensitive units, whereas the cathodal response in weak saccharin solutions reflected responses of sugar-sensitive units. To further test this assumption, we applied a crude decoction of *Gymnema sylvestre* and flooded the flow chamber with this substance for 1 min. In man, chewing the leaves of this plant is well known to block sweet sensitivity differentially and either saccharin or sugar becomes tasteless, so that sugar taken by mouth feels rather like sand in the mouth. The drug effect may last for ½ hr or more and was shown to block the electrophysiological responses of the human chorda tympani to sucrose and to saccharin (Diamant et al. 1963). The decoction of gymnemic acid in our experiments blocked the cathodal response to the sodium saccharin, but had no effect on the anodal response to the sodium saccharin, clearly involving sweet receptors in the cathodal stimulation via saccharin.

In our lab Dr. Pritchard carried out an analysis of single unit activity (Pfaffmann and Pritchard 1980). He tested units for their sensitivity to sodium chloride on the tongue as a chemical and their responsiveness to anodal stimulation via weak sodium chloride or sodium saccharin solutions. NaCl-reactive units clearly responded to the anode when the tongue was immersed in these subthreshold solutions, but they did not respond to the "cathode-on." Sugar-responsive single units responded to the cathode-on via a 1 mM sodium saccharin solution, but did not respond to the anode as shown in Figure 5.6. In the case of a few units that responded both to sodium chloride and to sucrose on the tongue, both the make of the anode and of the cathode were stimulating when polarized by way of subthreshold sodium saccharin solutions. In short, the responsiveness during polarization with weak anodal and cathodal currents matches the chemical sensitivity, that is, the sensitivity of the end organs with which these particular individual taste afferent fibers are associated.

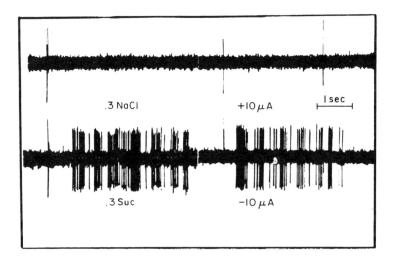

Figure 5.6. Four recordings from a single sugar sensitive hamster taste unit. The upper two traces show no response to 0.3 M NaCl or to + 10 μA; the lower two traces show responses to both 0.3 M sucrose and -10 μA via 0.001 M Na saccharin. This unit did not respond to solutions of NaCl, HCl, or quinine. (Reproduced with permission from Pfaffmann C, Pritchard T. In: van der Starre H, ed. Olfaction and Taste VII. IRL Press, London, 1980.)

Conclusion

The findings just described indicate that the first stimulatory step in electric taste at and just above threshold of activation is essentially chemical in nature. The weak electric currents are moving ions, excitatory cations or anions, as the case may be, into contact with or activation of the chemoreceptor surface. Thus, anodal currents activate taste receptors that are cation-sensitive. The relative efficiency of such cations is the same as their efficacy in solution at higher concentrations with no polarization. Cathodal currents activate receptor elements (receptor sites) reactive to the saccharin anions in solution or iontophoresed to the receptor surface. These sites are also reactive to sugar molecules. Levels of current flow higher than those we have used may involve other second-order effects.

I should like to make a further point in relation to the "psychophysiology of taste." In our experimental program we have interdigitated both behavioral methodology and strictly biophysical measurements of nerve action potentials. Either one of these methods can be a domain in its own right, but I have tried to show how the interaction or integration of diverse methods, those involving behavioral responses and those involving physiological measures, can lead to a solution that has, at least for me, more meaning and understanding not only of the mechanisms but of the function of receptors in the behaving organism.

Dr. Vincent Dethier of course has done essentially that in his studies of invertebrate chemoreception. So it is not necessary to argue the validity of this approach in this setting today, but merely to show that it can apply for warm-blooded creatures as well as to the invertebrates that have been the subject of Dr. Dethier's investigation. I would even suggest that the success in his hands has to some degree served as a model to those of us dealing with warmer-blooded, if not more complicated creatures than those with small brains. At the very least, ours are bigger and to some degree in the evolutionary sequence from rat to primate, they impinge a little more closely upon us as individuals.

Acknowledgment The research discussed herein was supported in part by a grant from the National Science Foundation (BNS 8111816).

References

Bujas Z (1971) Electrical taste. In: Beidler LM (ed) Handbook of Sensory Physiology Vol IV. Chemical Senses 2, Taste. Springer-Verlag, New York, pp 180–199

Bujas Z, Frank M, Pfaffmann C (1979) Neural effects of electrical taste stimuli. Sensory Proc 3:353–365

Carpenter JA (1956) Species differences in taste preferences. J Comp Physiol Psychol 49:139–144

Diamant H, Funakoshi M, Strom L, Zotterman Y (1963) Electrophysiological studies of human taste nerves. In: Zotterman Y (ed) Olfaction and Taste I. Pergamon Press, Oxford, pp 193–204

Fons M (1976) Electrically evoked taste threshold. Ann Otol Rhinol Laryngol 85:359–367

Frank ME, Hettinger TP, Herness MS, Pfaffmann CP (1986) Evaluation of taste function by electrogustometry. In: Meiselman HL, Rivlin RS (eds) Clinical Measurement of Taste and Smell. Collamore Press, Boston

Herness MS (1985) Neurophysiological and biophysical evidence of the mechanism of electric taste. J Gen Physiol 86:59–87

Hughes G (1969) Changes in taste sensitivity with advancing age. Gerontol Clin 11:224–230

Kashiwayanagi M, Yoshi K, Kobataki Y, Kurihara K (1981) Taste transduction mechanism. Similar effects of various modifications of gustatory receptors on neural responses to chemical and electrical stimulation in the frog. J Gen Physiol 78:259–275

Krarup B (1958) Electro-gustometry: a method for clinical taste examinations. Acta Otolaryngol 49:294–305

Moore LM, Nielsen CR, Mistretta CM (1982) Sucrose taste thresholds: age-related differences. J. Gerontol 37:64–69

Ninomiya Y, Funakoshi M (1980) Responses of the rat chorda tympani fibers to electric current of varying rate of rise applied to the tongue. In: van der Starre H (ed) Olfaction and Taste VII. IRL Press, London, p 217

Ninomiya Y, Funakoshi M (1981a) Responses of rat chorda tympani fibers to electrical stimulations of the tongue. Jpn J Physiol 31:559–570

Ninomiya Y, Funakoshi M (1981b) Role of ions in generation of taste nerve responses to electrical tongue stimulations in rats. Jpn J Physiol 31:891–902

Nowlis GH, Frank ME, Pfaffmann CP (1980) Specificity of acquired aversions to taste qualities in hamsters and rats. J. Comp Physiol Psychol 94:932–942

Oehrwall H (1891) Untersuchungen über der Geschmacksinn. Skand Arch Physiol 2:1–69

Pfaffmann C (1935) An experimental comparison of the method of single stimuli and the method of constant stimuli in gustation. Am J Psychol 48:470–476

Pfaffmann C (1941) Gustatory afferent impulses. J Cell Comp Physiol 17:243–258

Pfaffmann C (1985) De gustibus: praeterit, praesentis, futuri. In: Pfaff DW (ed) Taste, Olfaction and the Central Nervous System. Rockefeller University Press, New York, pp 19–44

Pfaffmann C, Pritchard T (1980) Ion specificity of "electric taste." In: van der Starre H (ed) Olfaction and Taste VII. IRL Press, London, pp 175–178

Smith D, Bealer SL (1975) Sensitivity of the rat gustatory system to the rate of stimulus onset. Physiol Behav 15:303–314

Volta A (1792) Briefe über thierische Electricitat. In: Oettingen AJ (ed) Ostwald's Klasiker der exakten Wissenschaften. Englemann, Leipzig. 1900. Quoted by Bujas Z (1971) Electrical taste. In: Beidler LM (ed) Handbook of Sensory Physiology IV. Chemical Senses 2, Taste. Springer-Verlag, New York, pp 180–199

Weiffenbach JM, Baum BJ, Burghauser R (1982) Taste thresholds: quality specific variation with human aging. J Gerontol 37:372–377

Chapter 6
What Makes a Caterpillar Eat?
The Sensory Code Underlying Feeding Behavior

LOUIS M. SCHOONHOVEN*

I can't explain *myself*, I'm afraid, sir,'' said Alice, ''because I'm not myself, you see.'' ''I don't see,'' said the Caterpillar.

Lewis Carroll

Applied entomologists (Forsyth 1803) and curious naturalists (Dethier 1937) have long been puzzled by the finickiness that caterpillars show with regard to their food preferences. Food specialists deprived of their natural food would rather succumb than accept unfamiliar plants. More general feeders are aware of differences between the various plant species that are acceptable to them (Merz 1959; Dethier and Kuch 1971; Hanson 1976). Although the striking feeding habits of phytophages held the fascination of entomologists for decades, analysis of this behavior had to wait till more became known of the chemical composition of plants and the differences between species. When electrophysiological techniques became available, the way was opened to monitor sensory input to the insect's brain, which guides feeding behavior. It is now possible to determine the way in which the sequence of decisions, representing feeding behavior (Dawkins and Dawkins 1973), is governed by external stimuli. Caterpillars seem to be ideal insects for such analysis, since (1) they show a great variety of (species-specific) feeding preferences and (2) they have a remarkably limited number of chemoreceptor cells to translate a plant's chemical make-up into a neural message on which the brain can take a decision. This chapter will concentrate on the attempts to decipher the sensory code used by caterpillars to guide their feeding behavior.

The Chemosensory Equipment of Caterpillars

Behavioral studies, including ablation experiments and histological examinations, have revealed that all external chemoreceptors in caterpillars are located in the antennae, maxillae, and epipharynx (Figure 6.1). Each antenna bears

---
*Department of Entomology, Agricultural University, Wageningen, The Netherlands.

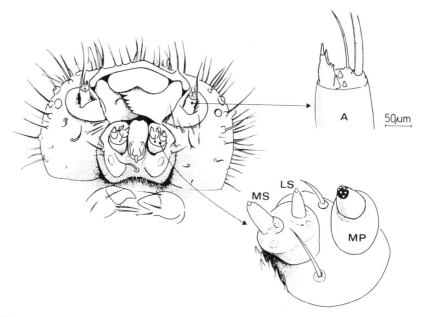

Figure 6.1. Ventral view of the head of a caterpillar (*Pieris brassicae*). A, antenna; MP, maxillary palp; LS and MS, lateral and medial sensilla styloconica.

three large sensilla basiconica, innervated by 16 olfactory neurons in total. Each maxilla has one palp and a galea, the latter carrying two sensilla styloconica, nonsocketed pegs with an apical papilla (Figure 6.2). These taste hairs are innervated by four bipolar neurons, the dendrites of which extend through the length of the hollow cuticular peg, ending just below the pore at the tip, i.e., within a few milliseconds of diffusion time from the external chemical world.

The tip of the maxillary palp is covered with eight sensilla basiconica. The pal

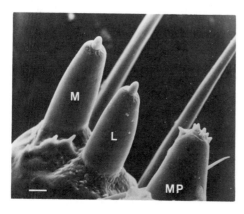

Figure 6.2. Scanning electron micrograph of lateral (L) and medial (M) sensilla styloconica on the maxilla of *Spodoptera littoralis* iarva. The maxillary palp (MP) is partly visible. Scale line: 5 μm. (Courtesy of W. M. Blaney, University of London.)

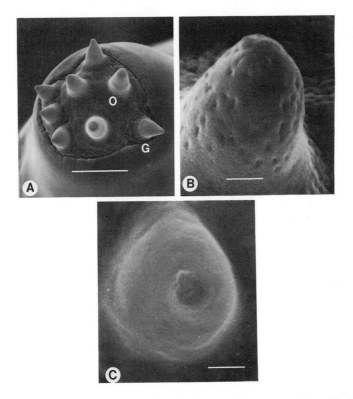

Figure 6.3. A. Tip of maxillary palp of *Heliothis zea*. Two types of sensilla basiconica are indicated by O (olfaction) and G (gustation). Scale line: 5 μm. B. Type O, a multiporous sensillum basiconicum that has a sculptured tip. Scale line: 0.5 μm. C. Type G, uniporous. Scale line: 0.5 μm. (Courtesy of D. A. Avé, Cornell University.)

tip receptors are innervated by 14–19 neurons in total (Schoonhoven and Dethier 1966). In *Euxoa messoria* they are innervated by 28 chemosensitive neurons (Devitt and Smith 1982). Scanning electron micrographs reveal that three of these sensilla basiconica are multiporous, and therefore presumably have an olfactory function, whereas the remaining ones are uniporous, typical of taste sensilla (Figure 6.3) (Dethier and Kuch 1971; Schoonhoven 1972; Hanson and Dethier 1973; Albert 1980; Avé 1981). In *E. messoria*, however, all eight sensilla appear to be of the uniporous type. In addition Devitt and Smith (1982) described the fine structure of two multiporous plate organs on the palps, equipped with three and four chemoreceptor neurons, respectively. One pair of gustatory organs is located on the epipharynx (Dethier 1937), innervated by three neurons (Ma 1972). These so-called epipharyngeal organs (Figure 6.4) are probably of general occurrence in lepidopterous larvae (Dethier 1975; Ma 1976; de Boer et al. 1977; van Drongelen 1979; Albert 1980; Avé 1981), although *Mamestra brassicae* and *E. messoria* seem to lack such sensilla (Blom 1978; Devitt and Smith 1982). Dethier (1937) attributed a gustatory function also to the hypopharynx,

Figure 6.4. Epipharyngeal taste organ of *Pieris brassicae*. Scale line: 1.2 μm. (From Ma W-C. Meded Landbouwhogesch Wageningen 72-11:1–162, 1972. Reprinted with permission.)

but was unable to find a putative chemosensory structure. Others have also occasionally obtained behavioral results that are difficult to understand without assuming the presence of some hitherto undetected gustatory organs in the preoral cavity (Ma 1976; Blom 1978).

As might be expected from their different positions, the four sets of receptors, i.e., antennal receptors, maxillary palp receptors, the two paired sensilla styloconica on the maxillae, and the epipharyngeal organs, serve different functions in the catenary process of food finding, food testing, and sustained feeding activity. From experiments in which these organs were inactivated or removed in various combinations, it is concluded that the sensilla styloconica and the epipharyngeal organs are the most critical receptors in the process of distinguishing host from nonhost plants (Torii and Morii 1948; Dethier 1953; Ito et al. 1959; Waldbauer and Fraenkel 1961; Waldbauer 1962, 1964; Schoonhoven and Dethier 1966; Ma 1972; de Boer et al. 1977; Blom 1978; Remorov 1982). When only the olfactory organs were ablated, the insect's ability to distinguish between different host plant species is diminished, but its host/nonhost discrimination is little affected (Hanson and Dethier 1973).

Although olfactory and gustatory sensilla often show marked microstructural differences, a sharp division between the two sensory modalities is sometimes difficult to make. For instance, the lateral sensilla styloconica on the maxillae undoubtedly have a gustatory function. Nonetheless, these sensilla have been found to respond also to volatiles emitted by damaged plant surfaces (Städler and Hanson 1975).

Olfactory Function of Antenna

From the results of an extensive series of behavioral experiments, including tests with caterpillars from which the antennae were extirpated, Dethier (1941) convincingly proved that the antennae have an olfactory function and are used in host plant recognition. Electrophysiological recordings from some olfactory cells in silkworm larvae supported this conclusion (Morita and Yamashita 1961). Experiments with a number of other lepidopterous larvae have shown that these receptors (or at least the majority of them) react as "generalists" and thus are sensitive to a wide variety of pure chemicals and plant volatiles. In an unstimulated situation they usually show some "spontaneous" activity, and stimulation may either decrease or increase the rate of firing. Moreover, each cell demonstrates a characteristic stimulus spectrum, and no two cells are identical, although their sensitivity spectra often overlap to a certain extent (Figure 6.5). Each volatile therefore stimulates several cells and is coded by activity patterns in a number of cells (Schoonhoven and Dethier 1966). Table 6.1 shows the responses of three cells to the volatiles of six host and two nonhost plants. Even when the ratios of response intensities are neglected, the odors of different plants produce different combinations of cell responses in most cases. It is most likely that, on the basis of differences between neural activity patterns, caterpillars are able to distinguish between the odors of many plant species,

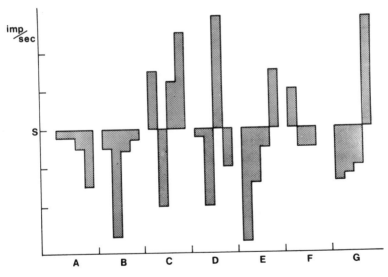

Figure 6.5. Activity profiles of four olfactory cells in the antenna of *Manduca sexta* during the first second of stimulation with seven different odors (A–G). S, Level of spontaneous activity. Inhibitory reactions occur more often than increases in firing rate. With the exception of one case (*odor F*), all cells react to all odors. (From Dethier VG. Am Physiol Soc Wash DC 1:79–96, 1967. Reprinted with permission.)

Table 6.1. Response patterns in three olfactory cells in the antenna of *Manduca sexta* larvae when stimulated with plant volatiles

	Stimulus	Cell number		
		1	2	3
Control:	Air	0	−	−
Host plants:	Tobacco	0	+	0
	Tomato	0	+	+
	Potato	−	+	0
	Black nightshade	+	0	0
	Bittersweet	+	0	0
	Solanum luteum	+	+	0
Nonhost plants:	Dandelion	+	+	+
	Carrot	−	+	0

+, −, 0: increase, decrease, and no change, respectively, of spontaneous activity.
Data from Schoonhoven LM, Dethier VG. Arch Neerl Zool 16:497–530, 1966.

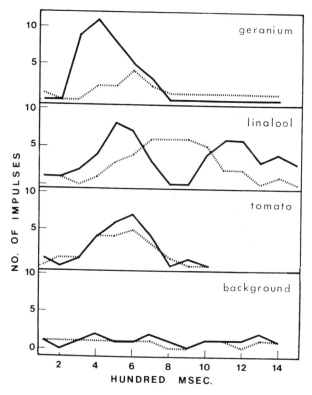

Figure 6.6. Response intensities of two olfactory cells in the antenna of *Manduca sexta* to three volatiles. (From Dethier VG, Schoonhoven LM. Entomol Exp Appl 12:535–543, 1969. Reprinted with permission.)

both acceptable and unacceptable (Dethier 1980b). A source of variation, other than mere increase or decrease of basal firing rates, is found when temporal relationships in firing activities between different olfactory cells are taken into account. For example, in *Manduca sexta* there are two cells that respond to the odor of tomato with a gradual increase of firing activity, reaching maximum activity at the same time and adapting at the same rate. Although these same cells react also to the odor of geranium, the temporal characteristics of the reactions of both cells are now very different (Figure 6.6). Such temporal patterning may add an extra dimension to the coding by olfactory receptors (Dethier and Schoonhoven 1969). To prove that the animal's brain is able to distinguish the differences between sensory patterns evoked by different odors, behavioral evidence is a prerequisite. Induction experiments have shown that caterpillars may develop a predilection for certain pure odors (Saxena and Schoonhoven 1978) or for plant species on which they have been reared. An odorous factor is involved also in this food choice behavior (Saxena and Schoonhoven 1982; Hanson 1983).

Although olfactory cells seem from the limited electrophysiological information presently available to be characterized by broad reaction spectra, it cannot be concluded that there are no cells in caterpillars with a narrow and/or very specific reaction spectrum. Indeed, J.H. Visser (personal communication) has found in the larva of *Pieris brassicae* an antennal receptor that specifically reacts to isothiocyanate and related compounds, substances that are produced by its host plants.

Chemoreception by Maxillary Palps

As mentioned before, the morphology of the palp tip sensilla indicates that some of them have an olfactory function, whereas the remainder are gustatory organs. In accordance with this observation, sensory responses to solutions as well as to odors have been demonstrated electrophysiologically. Like the antennal receptors, the olfactory cells of the palp possess broad response spectra (Schoonhoven and Dethier 1966). Hirao et al. (1976), recording from the maxillary nerves of the silkworm, observed that the spontaneous activity in some fibers is inhibited by the odor of two feeding stimulants, *n*-butyl proprionate and citral, but is increased when the palp is stimulated with peppermint oil, a repellent to this insect. Behavioral data support the idea that the maxillary palps are in some way involved in food recognition. Palpation of the intact leaf surface, prior to biting activity, may be related to contact chemoreception, during which chemicals on the leaf cuticle are perceived (Devitt and Smith 1985). Inactivation of the palps has the effect of eliminating the response of silkworms to the feeding stimulant *n*-butyl proprionate (Ishikawa et al. 1969) and impairs the ability of tobacco hornworms to recognize the host plant species on which they have been reared (Hanson and Dethier 1973). Interestingly, food intake on substrates that in intact caterpillars stimulate feeding only weakly or not at all increases considerably after inactivation of the palps. This is presumably due to the removal of spontaneous sensory activity, which in the CNS suppresses feeding activity (Ishikawa et al. 1969; Ma 1972; Blom 1978). The present evidence indicates that the maxillary palps participate in food recognition and that olfactory and contact chemoreceptors are involved in this process. But in spite of the fact that the palps house a considerable fraction of the caterpillars' total chemosensory equipment, it appears that their role is less prominent than that of the maxillary taste hairs, for example.

Sensory Coding of Taste

Because sense organs act as filters and relay information of only some specific conditions in the environment to the brain, they play a role in the decision process. This is not to say that the decision to eat or to reject a certain food is taken by the chemoreceptors of a caterpillar. But the way in which the chemical senses select information about the chemical consistency of plants (each plant of course comprises a myriad of compounds) reflects their degree of decisiveness (Dawkins and Dawkins 1973). It would be interesting to see whether it is possible to quantify the role of gustation in herbivorous caterpillars and to look for differences between species.

When a caterpillar bites into a leaf, the taste is coded by 11 (symmetric paired) receptors: four cells in each of the two maxillary sensilla styloconica and three cells in the epipharyngeal organ. Saps from different plants evoke different overall response patterns in these 11 cells, and the insect may show different behavioral reactions to these plants (e.g., Dethier and Crnjar 1982). The sensory code that underlies the behavioral responses may be unravelled by combining the analysis of sensory responses to (1) single chemicals, (2) mixtures, and (3) plant saps.

Responses to Single Chemicals

When the gustatory cells of a caterpillar are stimulated with solutions of a number of chemicals, it appears that different cells respond to different substances. For instance, the four cells of the medial taste hairs on the maxillae of *Pieris brassicae* are stimulated by respectively, salt, strychnine, sucrose, and sinalbin. The lateral hairs have cells for sucrose, sinigrin, proline, and anthocyanins, whereas the epipharyngeal organ cells respond to sucrose, strychnine, and salt. However, this does not mean that the gustatory cells are narrowly tuned to one compound only. The two strychnine cells react to a large variety of alkaloids and steroids, which all deter feeding in this species (Ma 1972). The sinigrin cell responds to many glucosinolates, the sinalbin cell to aromatic glucosinolates, and the proline cell to a number of amino acids (Schoonhoven 1967, 1969b). Thus, all cells possess a sensitivity spectrum, though one chemical, e.g., proline in the case of the amino acid cell, is the "best" stimulus for that cell; i.e., it evokes the strongest response. Moreover, some compounds may stimulate two or more cells, indicating that the response spectra of different cells are not strictly separated but show some overlap. Larvae of *Pieris brassicae* recognize their food essentially on the basis of the information from their eight maxillary taste cells because the epipharyngeal cells, duplicating some of the sensory capabilities of the maxillary taste hairs, seem not to add new information (Ma 1972, 1976; de Boer et al. 1977). Is it possible with our present knowledge of their taste cell characteristics to understand the food preferences of *Pieris brassicae* caterpillars? We will come back to this question later.

When trying to determine the degree of involvement of the chemical sense in decision-making in food selection, a comparison between species with different feeding habits may be helpful. If different species, having different food preferences, had identical chemoreceptors, it would have to be concluded that the CNS in different species take, on the basis of the same sensory message, different decisions. Consequently, the decisiveness of the chemical senses would be limited. However, if the sensory systems in different species were dissimilar, the contribution of the peripheral organs to the decision process would be evidently larger. Information on taste cell characteristics is available from more

Table 6.2. List of species in which the maxillary sensilla styloconica have been studied with electrophysiological techniques

Adoxophyes reticulana (15)	*Laothoe populi* (15)
Aglais urticae (15)	*Leucoma salicis* (15)
Bombyx mori (9, 10)	*Lymantria dispar* (7, 15)
Calpodes ethlius (7)	*Malacosoma americana* (5, 7)
Catocala nupta (15)	*Mamestra brassicae* (3, 18)
Celerio euphorbia (15)	*Manduca sexta* (15)
Ceratomia catalpae (7)	*Maruca testulalis* (14)
Chilo partellus (14)	*Mimas tiliae* (15)
Choristoneura fumiferana (1)	*Operophtera brumata* (15)
Cossus cossus (15)	*Papilio glaucus* (5)
Danaus plexippus (5, 6, 7)	*P. polyxenes* (5, 7)
Dendrolimus pini (11, 13, 15)	*P. troilus* (5)
Eldana saccharina (7)	*Pieris brassicae* (3, 11, 15)
Episema caeruleocephala (15)	*P. rapae* (7)
Estigmene acrea (7)	*Philosamia cynthia* (15)
Euchaetias egle (5)	*Pygarctia eglenensis* (5)
Heliothis armigera (17)	*Sphinx ligustri* (15)
H. virescens (17)	*Spodoptera exempta* (11, 12, 17)
H. zea (2, 7)	*S. littoralis* (17)
Isia isabella (7)	*Yponomeuta* (nine species) (8, 16)

References: (1) Albert 1980; (2) Avé 1981; (3) Blom 1978; (4) Clark 1980; (5) Dethier 1973; (6) Dethier 1980a; (7) Dethier and Kuch 1971; (8) van Drongelen 1979; (9) Ishikawa 1963; (10) Ishikawa 1966; (11) Ma 1972; (12) Ma 1977; (13) Menco et al. 1974; (14) den Otter and Kahoro 1983; (15) Schoonhoven 1972; (16) Schoonhoven et al. 1977; (17) Simmonds and Blaney 1984; (18) Wieczorek 1976.

Table 6.3. Receptor types found in various caterpillars

Sucrose (and other carbohydrates)	Amino acids
	Alkaloids, glycosides
Glucose	("deterrent cells")
Sorbitol	Glucosinolates
Inositol	Populin, salicin
Cation	Phlorizin
Anion	Adenosine
Water	Chlorogenic acid

than 40 lepidopterous species (Table 6.2). Although the data for most species must be considered as far from complete, some generalizations are possible.

The most striking feature when comparing the sensory characteristics of various lepidopterous larvae is that no two species have identical receptor systems (Dethier and Kuch 1971; Schoonhoven 1972). Apparently each species has evolved a (physiologically) unique chemosensory setup, tuned to recognizing

Table 6.4. Sensitivity spectra of sucrose receptors in the medial or lateral sensillum styloconicum

	P. brassicae Medial	M. brassicae Lateral	B. mori Lateral	D. pini Medial
Pentoses				
D-Arabinose	0	+	+	+
L-Arabinose	0	+	+	+
L-Rhamnose	NT	0	+	NT
D-Xylose	0	NT	+	+
Hexoses				
D-Fructose	+	+	+	+
D-Galactose	0	+	+	+
D-Glucose	+	+	+	+
D-Mannose	0	0	+	+
L-Sorbose	+	NT	+	+
Disaccharides				
Lactose	0	0	+	0
Maltose	0	+	+	0
Sucrose	+	+	+	+
Trehalose	0	0	+	NT
Cellobiose	NT	+	+	NT
Polyhydric alcohols				
Inositol	0	NT	0	+
Sorbitol	0	NT	+	NT

+ and 0, sensitive and insensitive; NT, not tested.
Data for *Pieris brassicae* and *Dendrolimus pini* after Ma (1972), for *Mamestra brassicae* after Wieczorek (1976), and for *Bombyx mori* after Ishikawa (1963).

its specific range of host plants in its particular biotope. Even related species
like *Pieris brassicae* and *P. rapae* (although some authors put *rapae* in the
genus *Artogeia*), both feeding on Cruciferae, show differences. Thus *P. rapae*
has receptors that are sensitive to salicin and inositol (Dethier and Kuch 1971),
whereas *P. brassicae* is unresponsive to these substances (Ma 1972). A group
of very closely related species of the genus *Yponomeuta*, each species living
on a different host plant, also shows marked differences in taste cell types
(Schoonhoven et al. 1977; van Drongelen 1979). Table 6.3 lists known cell types,
according to their "best" stimulus. Some cell types are of general occurrence,
e.g., the sucrose cell, whereas others have been found only in some species
of insects. Many, though not all, caterpillars have one or more cells that react
to sucrose and some other carbohydrates. Yet not all sucrose cells of different
insect species possess identical reaction spectra, as can be seen in Table 6.4.
Moreover, some species have, in addition to a broad-spectrum sugar receptor,
a cell type that responds, for example, only to glucose and fructose, or to inositol. Sorbitol receptors occur in species that live on host plants in which
sorbitol is the main carbohydrate in solution. The deterrent receptors, present
in many caterpillar species, show an even larger variability in response spectra
(Schoonhoven 1981).

It can be concluded that the taste hairs have receptors with which nutritive,
but also nonnutritive, compounds in plants can be detected. Although a number
of natural compounds stimulate two or, occasionally, more (see e.g., Dethier
1980b) cells, indicating some overlap, these receptors in general show, when
stimulated with *single* compounds, distinct response spectra, which we will call
their "overt spectrum" (see Figure 6.7A). The "overt spectrum" of the "su-
crose-best" cell includes sucrose and, depending on the species, several other
sugars. The fact that taste receptor properties vary widely between species,
together with the fact that each species has its own particular preferences, im-

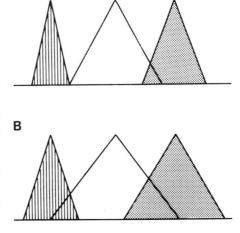

Figure 6.7. A. "Overt spectra" of three
different chemoreceptors. The broad-
spectrum cell in the middle shows some
overlap in responses to single compounds
with the receptor at the right. B. "Latent
spectra" of the same receptors. When
stimulated with mixtures the cells show
wider response spectra than can be concluded from stimulations with single
chemicals, and overlap is increased.

plies that the decision to accept a certain food or not is, to a certain extent at least, located in the sensory system.

Responses to Mixtures

When the stimulus solution comprises two chemicals, the receptor responses evoked often indicate some kind of interaction between them. Since no interaction between one chemosensory neuron and another has so far been found in insects, interactions between chemicals probably occur at the level of the dendritic membrane. Such interactions express themselves by increased or decreased responses as compared to stimulations with the same compounds singly. In the first paper published on taste cell responses in caterpillars Ishikawa (1963) presented several examples of interactions between chemicals, and since then this phenomenon has been demonstrated repeatedly by other workers in this field.

Interactions may occur between chemicals that, when applied singly, stimulate different receptor cells or, alternatively, act on the same cell. For example, the sinigrin-sensitive cell in the polyphagous *Isia isabella* larva is synergized by sucrose, which when applied singly stimulates only the sugar cell (Figure 6.8). This differs from the case in which two compounds each stimulate the same cell, and may in combination evoke an increased reaction compared with their single reactions (Table 6.5). The latter type of interactions of two adequate stimuli may be explained by assuming the presence of two or more different receptor sites in the receptor membrane with different stimulus-binding characteristics (Schoonhoven 1972; Menco et al. 1974; Wieczorek 1976).

Inhibitory effects of mixtures are also common. Some compounds, which behaviorally show antifeedant properties, derive this quality from their blocking effects on sugar receptors and thereby diminish the insect's capacity to perceive

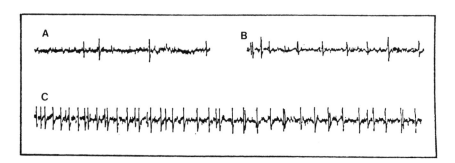

Figure 6.8. Synergistic receptor responses in medial sensillum of *Isia isabella*. A. Response to 0.001 *M* sinigrin. B. Response to 0.1 *M* sucrose. C. Response to a mixture of sinigrin and sucrose. The cell that preferentially responds to sinigrin alone shows a greatly increased response. (From Dethier VG, Kuch JH. Z Vgl Physiol 72:343–363, 1971. Reprinted with permission.)

Table 6.5. Impulse frequencies when stimulating the medial sensillum styloconicum of *Dendrolimus pini* with two carbohydrates and their mixtures at various concentrations

Stimulus	Impulse frequency action potentials·sec⁻¹
A (β-D-glucose at $1.6 \times 10^{-3}\ M$)	55
B (*meso*-inositol at $2.5 \times 10^{-4}\ M$)	27
2A	70
2B	39
A + B	111
2A + 2B	129

From Menco B et al. Proc K Ned Akad Wet C 77:157–170, 1974.

potent phagostimulants. Certain alkaloids, for example, suppress the sensitivity of the sugar-sensitive cells to sucrose in the red turnip beetle larvae (Mitchell and Sutcliffe 1984), and azadirachtin, a triterpenoid with antifeeding properties, reduces the sensitivity of the sugar cells in a number of caterpillars. At the same time sucrose may inhibit the response of the deterrent cells to azadirachtin, and thus counteract to a certain degree the diminished response of the sugar cell (Simmonds and Blaney 1984). Thus when mixtures are tested on a receptor it often appears that components that on their own are ineffective, stimulate or inhibit the response to the adequate stimulus. The sugar cell is modulated by salt (which may at low concentrations stimulate and at high concentrations inhibit the sugar response) and by, for example, azadirachtin. These compounds may be called "latent" stimuli of the sugar receptor since their effects on a receptor manifest themselves only in the presence of an adequate stimulus. It will be evident that the "latent spectrum" of a receptor may be considerably wider than its "overt spectrum" (Figure 6.7B). As a consequence receptors will in a natural situation, that is, when contacting complex chemical mixtures (plant saps), usually show their "latent spectra," and thus appear to be affected by more compounds than in the laboratory situation when tested with single compounds. Behavioral studies also show that chemical mixtures often evoke reactions that differ from the reactions to single compounds (e.g., Ishikawa et al. 1969; Meisner et al. 1972), thus confirming the neurophysiological findings. But of course such effects observed in behavioral studies are not necessarily due to peripheral interactions; in the course of processing sensory input in the CNS several types of interactions may occur additionally.

All available information indicates that interaction at the receptor level is a very common phenomenon. The types and degrees of such interactions represent an essential element in the coding of complex stimuli. The principles that underlie stimulus interactions are still wholly unknown, and the predictability of the phenomenon in specific instances is therefore close to naught. Interactions between salt and sugar, affecting the receptors for these compounds, have been found to exist in several instances (e.g., Ma 1972) and are probably of general occurrence. Their importance in quantitative terms, however, may vary from species to species.

It seems logical to assume that stimulus interaction in the sensory coding of natural substances has been evolved in order to increase an insect's discriminatory capacity with regard to food selection. Conceivably this is especially important in the case of caterpillars, which have only a very small number of receptors. Although the phenomenon of stimulus interaction has also been observed in locusts, for example, which have a multitude of chemoreceptors (Winstanley and Blaney 1978), it would be interesting to investigate whether the phenomenon is more important in insects with few receptors as compared to species having many of them.

Differences in feeding habits, for example between polyphagy and monophagy, could also determine the extent to which stimulus interaction in a given species occurs. Simmonds and Blaney (1984) found in a comparative study of nine species that an inhibitory effect of the feeding deterrent azadirachtin on the sugar cell responses occurs more often in polyphagous than in oligophagous species. But several interactions have also been described in the silkworm, a well-known food specialist. More information therefore is required before generalizations about a possible relationship between feeding habits and the occurrence of stimulus interactions at the receptor level can be made.

Responses to Plant Saps

Since a plant's sap is the natural stimulus for the maxillary sensilla styloconica and the epipharyngeal organ, saps freshly expressed from host or nonhost plants are the obvious stimuli to test on the taste receptors. Probably because the results obtained with such stimuli are difficult to interpret, only few studies of sensory reactions to plant saps have been published. When the two maxillary taste hairs of the tobacco hornworm were stimulated with saps from host and nonhost plants, two to four cells were activated in each sensillum. In view of the great diversity of plants tested this means that at least several of the receptors are not highly specific to compounds that are unique or typical of certain plant families. Different plant saps evoked different activity patterns, although it seemed difficult to relate the response characteristics to the insect's behavioral responses to these plants in terms of their acceptability as food (Schoonhoven and Dethier 1966).

Saps from a host plant and a nonhost stimulated all three cells present in the epipharyngeal organs of the same insect (de Boer et al. 1977). The most extensive study investigating the relationships between sensory input from plant saps and food acceptance behavior was done by Dethier (1973) using some monophagous as well as some oligophagous species. It confirmed the conclusion that different plant saps, either from acceptable or unacceptable plants, give different responses. Furthermore it was found that the sensory response patterns to the same plant saps are different in different caterpillar species (this agrees with the findings that different species possess different receptor specificities, as discussed before). An extension of these conclusions leads to the dictum that, since there is no single standard electrophysiological response to rejected plants, rejection is not a unitary modality, nor is there a universal difference between sensory patterns for acceptance and those for rejection. The concept of across-fiber patterning, in which each plant's unique sensory response pattern maintains its characteristic features, seems to fit best all the evidence available (Dethier 1973). It may also explain the curious finding that an antifeedant presented to caterpillars of *Spodoptera littoralis* on their preferred food (cabbage) is much more effective than when offered on a less preferred food (wheat) (Simmonds et al. 1985). The total sensory pattern evoked by the combination of cabbage sap and antifeedant is, possibly due to certain interactions, less acceptable than that from wheat and antifeedants.

CNS Interpretation of Sensory Code

Feeding activity requires motor output from the CNS. Chemosensory input may, depending on its message, stimulate or inhibit feeding motor output. In the complete absence of sensory input, i.e., after inactivation of all chemoreceptors, some insects may feed indiscriminately (de Boer et al. 1977) whereas others refuse to eat (Ma 1972). The process in which the CNS evaluates the sensory input, resulting in either continuation or cessation of feeding activity, is at present not accessible to direct study. For the time being we can only analyze the sensory input, and, by comparing this with behavioral responses, hypothesize about the principles that underlie the central processing system. The following points summarize some basic considerations. (1) The gustatory sense has the leading role in feeding activity. The epipharyngeal organs do not add new information to that of the maxillary hairs. (2) Sensory input from the maxillae is sent to the suboesophageal ganglion, that from the epipharyngeal organs to the tritocerebrum. (3) The message indicating whether a plant is acceptable or not must be hidden in the sensory pattern it evokes. If the CNS is able to read this message, it is in principle also decipherable to us. Different messages may be considered as different "keys," some of which may "unlock," i.e., permit feeding activity, others not.

Unravelling of the characteristics of the CNS processing system may be done along two lines. The first method is based on using simple stimuli and quantifying the relationships between simple and modifiable sensory inputs and feeding activity. Once the effects of single compounds are known, stimuli of gradually increasing complexity can be tested in the same way. Ultimately the synthesis of stimuli mimicking natural stimuli may be used. A second method starts from the other end and uses natural stimuli. By analyzing the complex sensory patterns they evoke and by correlating certain characteristics with feeding behavior, properties of the central recognition process, the "lock," may be inferred. Both methods, of course, have their limitations, but a combination of the insights gained by each of them may unveil elementary principles of CNS processes.

Single-Line Codes

This method is based on our knowledge of certain chemicals stimulating particular cells only. When adding such chemicals to a neutral substrate, such as an agar-cellulose diet, exact quantitative relationships can be obtained between impulse frequencies, known to be evoked by these chemicals at the concentrations applied, and the amount of diet intake. Such studies have been made with caterpillars of *Pieris brassicae* (Ma 1972; Blom 1978). It appears that very strong correlations exist between numbers of nerve impulses evoked by certain chemicals at certain concentrations and feeding intensity (expressed as amounts of fecal pellets produced per 24 hr) induced by them (Figure 6.9). For seven out of 11 gustatory cells present the number of receptor impulses equivalent to a standard feeding intensity can be calculated (Table 6.6). The data at present available agree well with the assumption that impulse trains entering the CNS via different lines are combined according to simple arithmetic laws, i.e., addition and subtraction, and its outcome is reflected directly in the amount of diet eaten. Impulses from various lines, which carry information on feeding stimulants, have more or less the same weight, but impulses from deterrent

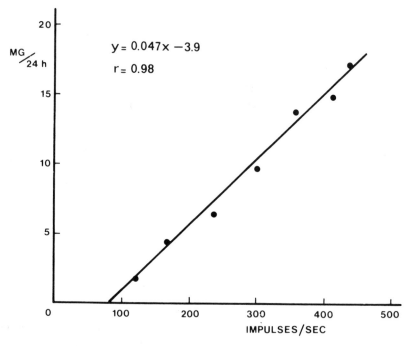

Figure 6.9. Relationship between receptor activity (sum of action potentials in lateral and medial sensilla basiconica and epipharyngeal organ in first second after stimulation) in response to various sugar concentrations and food intake in *Pieris brassicae* (expressed as mg dry weight of fecal pellets per 24 hr) in response to various sugar levels in an agar-cellulose medium. (Data from Blom F. Neth J Zool 28:277–340, 1978. Reprinted with permission.)

Table 6.6. Number of receptor impulses per
second above a certain threshold required to
induce (+) or reduce (−) a feeding intensity
equivalent to 1 mg fecal pellets/24 hr

+ Sugar cells:	21 (+ threshold = 81)
+ Amino acid cell:	15
+ Glucosinolate cell:	18
− Deterrent cells:	8 (+ threshold = 62)

cells appear to have a value about 2½ times higher (Table 6.6). The quantitative
relationship between sensory input from seven receptors and feeding activity
is depicted in a model (Figure 6.10). There can be no doubt that the inputs from
the different chemoreceptors, although small in number in caterpillars, must
converge somewhere in the CNS, when the decision "to eat or not to eat" can
indeed be classified as an all-or-none type of behavior (Bullock 1961). The feeding center (FC in Figure 6.10), possibly even represented by a single neuron

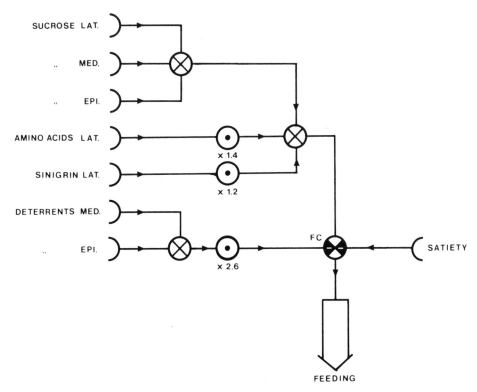

Figure 6.10. Diagram of CNS processing neural input from seven chemoreceptors and
receptors that signal degree of satiety. Relative weight of receptors for different stimulus
qualities is given as a multiplication factor with sugar cells as standard, i.e., each sugar
cell = 1.

(Nolen and Hoy 1984), which lies at the head of a hierarchy of motor commands, is the ultimate integrator of afferent inputs.

The high correlations between behavioral and neurophysiological data seem to prove conclusively that the CNS is able to read quantitative variations in sensory input in a highly accurate way, an intuitive deduction reached before from observations on food selection.

Synergistic behavioral reactions to mixtures of two (or more) chemicals may be due, as discussed above, to synergistic interactions at the receptor level or at higher integration levels, i.e., during the processing of sensory information. Although several instances of synergism at the behavioral level are known (e.g., Thorsteinson 1960; Ishikawa et al. 1969), it is as yet unknown if any of them results from synergistic processes at the central level.

Multiple-Line Codes

Natural stimuli, such as plant saps, often fire three to four cells in each taste hair, though now and then only two or even one cell in a sensillum is active (Schoonhoven and Dethier 1966; Dethier 1973). Since it is unknown which features of the sensory message cause the CNS to accept or to reject a plant when the insect bites it, it is logical to look for variables in the code that correlate with acceptability.

A gross method is to compare the total impulse frequency of the medial hair to that of the lateral hair after stimulation by the same plant sap. It appears that in tobacco hornworms acceptable plants in most cases evoke higher impulse frequencies in the medial hair than in the lateral one, whereas unacceptable plant species show a reverse frequency ratio (Figure 6.11) (Schoonhoven and Dethier 1966; Schoonhoven 1969a).

Dethier and Crnjar (1982), in an elegant analysis of impulse patterns in the tobacco hornworm, detected several differences between sensory responses to saps from different plants. They concluded that during the initial phasic response period only the salt receptor fires, though its impulse frequency and the time of its maximal activity may vary with plant species. During the subsequent tonic phase this cell showed different spike interval distributions on stimulation with saps from different host plants. It is known that different cells may differ in their spike interval distributions (e.g., Blom 1978), but the above observation indicates that the nature of the stimulus may affect the temporal response characteristics of one and the same receptor cell. Moreover it was found not only that the regularity of firing may vary, but also that the distribution of spike intervals is not necessarily random, depending on stimulus type. Thus, by measuring the higher-order interspike intervals, it appears that different temporal patterns in the responses of receptors exist, which vary with the different saps.

When host plant saps were tested, the relative firing ratios of three cells were different for Jerusalem cherry compared with tobacco and tomato. This ensemble firing ratio represents another coding principle that, in addition to the temporal response characteristics of the phasic period and the spike interval

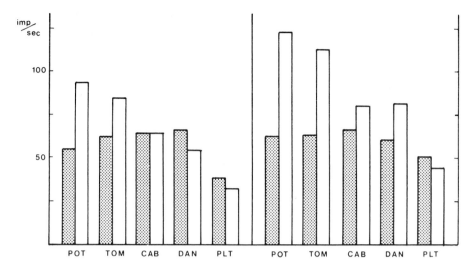

Figure 6.11. Impulse frequencies in lateral (dotted columns) and medial (open columns) sensilla styloconica in *Manduca sexta* larvae after stimulation with saps from potato (POT), tomato (TOM), cabbage (CAB), dandelion (DAN), and plantain (PLT). The left side of the figure refers to larvae reared on tomato (these insects refuse to eat cabbage, dandelion, and plantain); the right side of the figure refers to larvae reared on an artificial diet (these insects readily eat cabbage, plantain and, with some reluctance, dandelion). (After data from Schoonhoven LM. Proc K Ned Akad Wet 72:491–498, 1969.)

patterning in the tonic period, might be used by the CNS. It remains to be determined, however, which of these or other hitherto undetected variables in the code are read by the CNS.

In the foregoing, two approaches to elucidate the nature of the sensory code have been described. One method employs sharply defined single cell responses whereas the more holistic mode searches for patterns in ensemble firing activities. At first glance it might seem that the first method is linked to the concept of "labelled lines," and that the second way must lead to adoption of the theory of "across-fiber patterning." These two concepts are often considered as contrary and mutually exclusive. The debate as to which of the two theories best explains the existing information on taste in vertebrates has been going on for many years and probably will not be settled soon (see e.g., Scott and Chang 1984).

In view of our present knowledge it seems unfruitful when trying to understand sensory coding in insects to emphasize the antagonistic position of the two theories. The two concepts, when not used in their extreme denotations, may both help to explain the situation in caterpillars (see also Städler 1982).

Some cells in a given species may have very narrow reaction spectra (such as possibly some inositol cells and water cells), others have a broader spectrum, though still with little overlap with others (e.g., one of the two glucosinolate cells in *Pieris brassicae*), whereas a third category of cells, having broad "latent spectra," may seem ideal for participation in an across-fiber patterning. Complex stimuli will evoke a complex impulse pattern from a mixture of narrowly tuned and broad-spectrum receptors, showing all the features of an across-fiber pattern. Other, also complex, stimuli may in the same insect activate only one or two cells (not too rare in caterpillars), and features of the labelled-line concept may appear. Dethier and Crnjar (1982) aptly remark that an insect's receptor system may act as a labelled-line system in one situation (for example when stimulated by one chemical) and as an across-fiber pattern system in another. In conclusion the sensory coding in the taste receptors of caterpillars may show, depending on stimulus type and insect species, the combined characteristics of the two coding principles, i.e., the labelled-line concept and across-fiber patterning.

Sensory Code and Time

The CNS needs a minimum integration time when assessing across-fiber patterns of codes based on temporal characteristics. Whereas behavioral reaction times to deterrent solutions can be as short as 200 msec (in this case we are usually dealing with a simple "labelled-line" response), discrimination between two acceptable host plants requires, in the tobacco hornworm at least, 5 sec (Dethier and Crnjar 1982). It is reasonable to suppose that it takes a longer time to determine differences, when they are of a more subtle nature. The blowfly, for example, needs 0.5–1.0 sec to integrate sensory input in order to discriminate different concentrations of sucrose (Smith et al. 1984).

Once a caterpillar has decided to accept a food, feeding activity continues as if it possesses a momentum. This is well illustrated by larvae of *Danaus plexippus*, which, when feeding on their normal host plant, will take an average of seven bites of an otherwise strictly unacceptable plant, when this is suddenly held in front of its mouthparts (Mayer and Soule 1906).

Another aspect of time in relation to coding relates to sensory adaptation. When different receptors adapt at different rates and reach different levels relative to their initial response intensities, the sensory code will change with time. There are indications that the deterrent receptor in *Pieris brassicae* shows adaptation characteristics differing from those of some other receptors (Schoonhoven 1977). This means that in the course of a meal that initially is sufficiently phagostimulating to start feeding, the contribution of the deterrent receptors may become more preponderant, resulting in an early cessation of the meal as compared to normal.

Summary of Our Understanding of the Sensory Code

The results of studies aimed at elucidating the sensory code used by caterpillars in food selection may be summarized as follows:

1. The code is different for each insect species.
2. Synergistic and inhibitory interactions at the peripheral level are common.

3. Response intensities in specific receptors show (in *Pieris brassicae*) strong correlations with behavioral reactions. Central processing seems to be based on simple arithmetic rules. Impulses from different receptors may be valued (in quantitative terms) differently.
4. A plant's chemical fingerprint is coded by eight receptors, using across-fiber patterning. This code is most probably based on multidimensional firing ratios. It may include temporal characteristics of receptor responses.

Sensory Code and Food Plant Range

The data available at present suggest that in those species that have been investigated "there is no universal difference between sensory patterns for acceptance and those for rejection" (Dethier 1973), although the finding that in *Manduca sexta* the firing ratios of the two maxillary taste hairs are correlated with plant acceptability may seem to contradict this conclusion. The sensory pattern of a specialist feeder apparently would have to match more closely a certain norm set by the CNS, in order to trigger feeding activity, than is the case in generalist feeders. In other words, many different receptor activity profiles or "keys" fit into the CNS template ("lock") and release feeding in generalists, whereas the "locks" of specialists are more selective (Figure 6.12). The decision whether or not a certain plant will be eaten thus depends on two elements: the sensory pattern on one hand (which is the product of receptor characteristic) and the weighing process in the CNS on the other hand. The decision process requires a close cooperation between both sides and the decisiveness of the system (Dawkins and Dawkins 1973) pertains to the whole, rather than to one of its elements. The key-lock model, although a metaphor, can be tested experimentally and therefore could help to guide our thinking of central processing principles.

Students of sensory patterns in caterpillars are often confronted with a great variability in sensory responses, which cannot be eliminated by rigid standardization of experimental procedures (Schoonhoven 1976, 1977; van der Molen et al. 1985). Since also behavioral variations occur, preferably each individual should be used as its own control as far as possible and electrophysiological and behavioral observations should preferably be made on the same individuals. Especially in insect species with a narrow food range, variations in their sensory

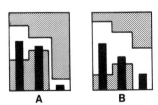

A   **B**

Figure 6.12. Model of CNS processing of sensory input. The black bars represent action potential frequencies in three chemoreceptors when stimulated by an acceptable food plant. The white space of the "lock" reflects the variation allowed to the sensory input, in order to be interpreted as acceptable. (A) Food specialist. (B) Food generalist.

patterns should be compensated by corresponding alterations in their CNS. In cases where the variations at both sides, i.e., chemoreceptors and CNS, are not fully complementary to each other, individual variations in behavior arise. Such may be the case in certain individuals which without obvious hesitation relish certain nonhost plants (Merz 1959; Schoonhoven 1977).

Chemoreceptor sensitivity in a number of insect species has recently been found to vary to a certain extent with age, time of the day, and/or feeding history (Blaney et al. 1986). Although the changes observed are often relatively small, e.g., when compared with individual variation, they are likely to be noticed by the CNS. In some instances changes in sensory patterns occurred synchronously with behavioral changes, e.g., acceptance of certain nonhost plants as food (Figure 6.11) (Schoonhoven 1969a), or preferences developed for a particular host plant species (Städler and Hanson 1976), suggesting that behavior is modified by altered sensory input. In these cases where receptor sensitivity changes occur as a result of exposure to certain diets, learning (i.e., changes in behavior as a result of former experiences) seems to be based, at least partly, on changes in the peripheral nervous system.

Another factor that might affect the sensory code, and consequently its central interpretation, is temperature. When all chemoreceptors have identical temperature characteristics, no essential change in pattern will occur when the ambient temperature changes, and the CNS is essentially confronted with the same sensory code (provided the stimulus, i.e., the plant's chemical composition, does not change with temperature). However, when different taste cells have different temperature responses, as is the case in two chemoreceptors in setae on the legs of crayfish (Hatt 1983a,b), sensory profiles will vary with temperature. The CNS then will have to compensate for temperature effects, or behavioral responses will change with temperature.

Other Factors Affecting Feeding Activity

Chemosensory input is of course not the only factor that determines whether or not the CNS will switch on feeding activity; neither is switching it off only governed by satiety as measured by stretch receptors in the alimentary tract. Suboptimal food quality is reflected in shorter durations of a meal (Ma 1972). This suggests that in the CNS a continuous weighing process between chemosensory input from the food and satiety signals takes place. Sensory adaptation may play a role in this process as well.

The CNS is also aware of the physiological status of the insect (Figure 6.13), a general term that includes feedback from nutritional factors, nutritional deficiencies, and toxins. There are several pieces of evidence for the existence of such feedback. One example is found in polyphagous insects with a preference for mixed diets. Merz (1959) reported that caterpillars of *Malacosoma castrensis* and *Arctia caja* prefer to change their food from time to time and grow better when they are allowed to do so. Corn earworm larvae when offered two artificial diets, which complement each other in nutritional respects, eat such proportions

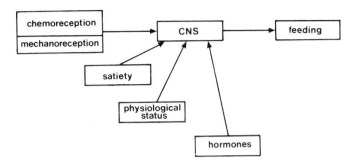

Figure 6.13. Factors that may affect the CNS center controlling initiation and cessation of feeding activity.

from each diet that a nutritionally balanced mixture is obtained (Waldbauer et al. 1984). These examples suggest that the norm set by the CNS, to which sensory input is gauged, is not a constant one, but rather can be adapted to the animal's nutritional needs.

Another example of physiological feedback is found in polyphagous caterpillars, which will feed on certain toxic plant species. After recovery from sickness due to the ingestion of toxins, the caterpillars will avoid eating this plant, which indicates again the presence of a physiological feedback system, in this case coupled to a learning process (Dethier 1980c). Likewise the decrease in food intake in caterpillars that are infected with *Bacillus thuringiensis* is probably due to an inhibition of their feeding centres by the bacterial endotoxin (Retnakaran et al. 1983).

The fact that food intake is interrupted some time before each moult suggests a hormonal influence on the feeding center in the CNS. It has indeed been reported that juvenile hormone stimulates food intake in lepidopterous larvae (Sieber and Benz 1978; Dominick and Truman 1984; Muraleedharan and Prabhu 1981). The observation that several caterpillar species during the first day after a moult are much more finicky in their food choice than during the remaining days of the same instar (Merz 1959) may also be due to a hormonal effect on the CNS.

Thus the system that in the CNS evaluates incoming sensory codes, the "lock," cannot be considered as a fixed and constant gate, but rather shows some versatility by its responses to signals reflecting the animal's physiological condition and its hormonal situation. In addition, learning processes located within the CNS may modify the properties of this evaluation center, which express themselves in changed feeding behavior.

Evolutionary Aspects

The evolution of food preferences in phytophagous insects is generally thought to be strongly influenced by the development of secondary plant substances in the plant world. The evolutionary routes that have led to the present-day situation, however, are difficult to trace and therefore largely a matter of conjecture (see e.g., Dethier 1954; Jermy 1983, 1984). The question arises whether sensory physiology can contribute to our understanding of how the present insect–plant relationships have evolved. As pointed out earlier, chemoreceptor characteristics vary widely between species and there seem to be no two species, even when closely related to each other, with an identical chemosensory system. This has led to the assumption (Schoonhoven et al. 1977; Schoonhoven 1981) that chemoreceptor characteristics can easily be modified and adapted to new situations, when an insect changes its food-plant range. It would be interesting to compare the sensory characteristics of individuals of different populations, especially when these populations are found on different food plants, for example as described for *Saturnia pavonia* (Merz 1959).

Some cases of genetic differences of receptor characteristics are known in insects, for instance, in the deterrent receptor of two strains of *Mamestra brassicae* (Wieczorek 1976). Interestingly two races of *Yponomeuta padellus*, living on different hosts, have identical gustatory responses, although it cannot be excluded that subtle differences have escaped attention (van Drongelen 1979). The genetics of chemoreception is open for research, as was shown by van Drongelen and van Loon (1980). They crossed two *Yponomeuta* species and found that the hybrid larvae were fully sensitive to two chemicals that were stimulating to only one of the parents, whereas the sensitivity to a third compound showed an intermediate expression. The hybrids accepted the host plants of both parents. More studies of this type are needed, since they are essential for a better understanding of the genetics of chemoreceptors in relation to feeding preferences.

Gustatory receptors may be classified into four types of cells sensitive to: nutrients, salts, phagostimulating allelochemics, and deterrents. Salt receptors probably represent a cell type that evolved early and is little changed from cells that may be designated as common chemical sense. It has been proposed (Dethier 1980a) that herbivorous insects have evolved, also from the common chemical sense, receptors sensitive to a wide variety of plant compounds. Some of these cells are deterrent receptors; others detect phagostimulating allelochemics.

Since sugar receptors occur universally in animals and are found even in bacteria, they have probably also a very long history. The fact that some deterrents inhibit sugar receptors, whereas sugar may interfere with deterrent cell activity (Simmonds and Blaney 1984), and that interactions commonly occur between salts and sugars and their respective receptors, suggests that our classification of chemoreceptors has a certain degree of arbitrariness.

As mentioned earlier, the variation of receptor types is immense and the phenomenon of interactions between different categories of stimuli has hardly been analyzed systematically, probably leaving another source of variation still to be disclosed. In view of this it seems questionable whether, on the basis of physiological knowledge of present-day chemoreceptors, conclusions on their evolution in insects can ever be obtained.

Conclusion

Lepidopterous larvae are characterized by an enormous diversity in the range of sensitivity of their chemoreceptors, perhaps unsurpassed in the animal world. As a consequence it appears that the question: What makes a caterpillar eat? is more difficult to answer than would be expected on the basis of its remarkably small number of chemoreceptors. On the other hand this finding shows the ingenuity of Nature under circumstances where organisms are confronted with a complex chemical world, and at the same time it provides students of chemoreception, investigating probably the most basic sense, with a rich source of materials.

Questions such as that posed in the title of this chapter may not only lead to the discovery of fundamental principles underlying the interactions between animals and plants, but may also help to find new means to control the largest group of pest insects on our crops. This idea, which was expressed by Forsyth as early as 1803, when he wrote about caterpillars and stated:

> It would be of great service to get acquainted as much as possible with the economy and natural history of all these insects, as we might thereby be enabled to find out the most certain method of destroying them,

together with the new insights gained so far, may be a justification for increasing our efforts to understand a caterpillar's feeding behavior. Fortunately, V.G. Dethier has provided this field with a broad and firm experimental basis and has enriched the subject with a wealth of fruitful concepts and stimulating thoughts.

Acknowledgments I thank Dr. W. M. Blaney and Dr. M. S. J. Simmonds for stimulating discussions.

References

Albert PJ (1980) Morphology and innervation of mouthpart sensilla in larvae of spruce budworm *Choristoneura fumiferana* (Clem.) (Lepidoptera: Tortricidae). Can J Zool 58:842–851

Avé DA (1981) Induction of changes in the gustatory responses by individual secondary plant compounds in larvae of *Heliothis zea* (Boddie) (Lepidoptera, Noctuidae). Ph.D. thesis Mississippi State University. Mississippi State, MI

Blaney WM, Schoonhoven LM, Simmonds MSJ (1986) Sensitivity variations in insect chemoreceptors; a review. Experientia 42:13–19

Blom F (1978) Sensory activity and food intake: a study of input-output relationships in two phytophagous insects. Neth J Zool 28:277–340

Bullock TH (1961) The problem of recognition in an analyser made of neurons. In: Rosenblith WA (ed) Sensory Communication, MIT Press, pp 717–724

Clark JV (1980) Changes in the feeding rate and receptor sensitivity over the last instar of the African armyworm, *Spodoptera exempta*. Entomol Exp Appl 27:144–148

Dawkins R, Dawkins M (1973) Decisions and the uncertainty of behaviour. Behaviour 45:83–103

de Boer G, Dethier VG, Schoonhoven LM (1977) Chemoreceptors in the preoral cavity of the tobacco hornworm, *Manduca sexta,* and their possible function in feeding behaviour. Entomol Exp Appl 21:287–298

den Otter CJ, Kahoro HM (1983) Taste cell responses of stem-borer larvae, *Chilo partellus* (Swinhoe), *Eldana saccharina* Wlk, and *Maruca testulalis* (Geyer) to plant substances. Insect Sci Appl 4:153–157

Dethier VG (1937) Gustation and olfaction in lepidopterous larvae. Biol Bull Woods Hole Mass 72:7–23

Dethier VG (1941) The function of the antennal receptors in lepidopterous larvae. Biol Bull Woods Hole Mass 80:403–414

Dethier VG (1953) Host plant reception in phytophagous insects. 9th Int Congr Entomol. Amsterdam 2:81–83

Dethier VG (1954) Evolution of feeding preferences in phytophagous insects. Evolution 8:33–54

Dethier VG (1967) Feeding and drinking behaviour of vertebrates. In: Code CF (ed) Handbook of physiology. B. Alimentary canal. Am Physiol Soc Washington DC 1:79–96

Dethier VG (1973) Electrophysiological studies of gustation in lepidopterous larvae. II. Taste spectra in relation to food-plant discrimination. J Comp Physiol 82:103–134

Dethier VG (1975) The monarch revisited. J Kans Entomol Soc 48:129–140

Dethier VG (1980a) Evolution of receptor sensitivity to secondary plant substances with special reference to deterrents. Am Natur 115:45–66

Dethier VG (1980b) Responses of some olfactory receptors of eastern tent caterpillar *(Malacosoma americanum)* to leaves. J Chem Ecol 6:213–220

Dethier VG (1980c) Food-aversion learning in two polyphagous caterpillars, *Diacrisia virginica* and *Estigmene congrua.* Physiol Entomol 5:321–325

Dethier VG, Crnjar RM (1982) Candidate codes in the gustatory system of caterpillars. J Gen Physiol 79:543–569

Dethier VG, Kuch JH (1971) Electrophysiological studies of gustation in lepidopterous larvae. I. Comparative sensitivity to sugars, amino acids, and glycosides. Z Vgl Physiol 72:343–363

Dethier VG, Schoonhoven LM (1969) Olfactory coding by lepidopterous larvae. Entomol Exp Appl 12:535–543

Devitt BD, Smith JJB (1982) Morphology and fine structure of mouthpart sensilla in the dark-sided cutworm *Euxoa messoria* (Harris) (Lepidoptera: Noctuidae). Int J Insect Morphol Embryol 11:255–270

Devitt BD, Smith JJB (1985) Action of mouthparts during feeding in the dark-sided cutworm, *Euxoa messoria* (Lepidoptera: Noctuidae). Can Entomol 117:343–349

Dominick OS, Truman JW (1984) The physiology of wandering behaviour in *Manduca sexta.* I. Temporal organization and the influence of the internal and external environments. J Exp Biol 110:35–41

Forsyth W (1803) A treatise on the culture and management of fruit-trees. 2nd Edition Longman, London

Hanson FE (1976) Comparative studies on induction of food choice preferences in lepidopterous larvae. Symp Biol Hung 16:71–77

Hanson FE (1983) The behavioral and neurophysiological basis of foodplant selection by lepidopterous larvae. In: Ahmad S (ed) Herbivorous Insects: Host Seeking Behavior and Mechanisms. Academic Press, New York, pp 3–23

Hanson FE, Dethier VG (1973) Role of gustation and olfaction in food plant discrimination in the tobacco hornworm, *Manduca sexta*. J Insect Physiol 19:1019–1034

Hatt H (1983a) Temperature dependence of the response of the pyridine-sensitive units in the crayfish walking leg. J Comp Physiol A 152:395–403

Hatt H (1983b) Effect of temperature on the activity of the amino acid receptors in the crayfish walking leg. J Comp Physiol A 152:405–409

Hirao T, Yamaoka K, Arai N (1976) Studies on mechanism of feeding in the silkworm, *Bombyx mori* L. II. Control of mandibular biting by olfactory information through maxillary sensilla basiconica. (in Japanese with English summary) Bull Seric Exp Stn Jpn 26:385–410

Ishikawa S (1963) Responses of maxillary chemoreceptors in the larva of the silkworm, *Bombyx mori*, to stimulation by carbohydrates. J Cell Comp Physiol 61:99–107

Ishikawa S (1966) Electrical response and function of a bitter substance receptor associated with the maxillary sensilla of the silkworm, *Bombyx mori* L. J Cell Physiol 67:1–11

Ishikawa S, Hirao T, Arai N (1969) Chemosensory basis of host plant selection in the silkworm. Entomol Exp Appl 12:544–545

Ito T, Hori Y, Fraenkel G (1959) Feeding on cabbage and cherry leaves by maxillectomized silkworm larvae. J Seric Sci 32:128–129

Jermy T (1983) On the Evolution of Insect/Host Plant Systems. Verh SIEEC X Budapest, pp 13–17

Jermy T (1984) Evolution of insect/host plant relationships. Am Natur 124:609–630

Ma W-C (1972) Dynamics of feeding responses in *Pieris brassicae* Linn. as a function of chemosensory input: a behavioural, ultrastructural and electrophysiological study. Meded Landbouwhogesch Wageningen 72-11:1–162

Ma W-C (1976) Mouth parts and receptors involved in feeding behaviour and sugar perception in the African armyworm *Spodoptera exempta* (Lepidoptera, Noctuidae). Symp Biol Hung 16:139–151

Ma W-C (1977) Electrophysiological evidence for chemosensitivity to adenosine, adenine and sugars in *Spodoptera exempta* and related species. Experientia 33:356–357

Mayer AG, Soule CG (1906) Some reactions of caterpillars and moths. J Exp Zool 3:415–433

Meisner J, Ascher JRS, Flowers HM (1972) The feeding response of the larva of the Egyptian cotton leafworm, *Spodoptera littoralis* Boisd. to sugars and related compounds. I. Phagostimulatory and deterrent effects. Comp Biochem Physiol A 42:899–914

Menco B, Schoonhoven LM, Visser J (1974) Qualitative and quantitative analysis of electrophysiological responses of an insect taste receptor. Proc K Ned Akad Wet C 77:157–170

Merz E (1959) Pflanzen und Raupen. Biol Zentralbl 78:152–188

Mitchell BK, Sutcliffe JF (1984) Sensory inhibition as a mechanism of feeding deterrence: effects of three alkaloids on leaf beetle feeding. Physiol Entomol 9:57–64

Morita H, Yamashita S (1961) Receptor potentials recorded from the sensilla basiconica on the antenna of the silkworm larva, *Bombyx mori*. J Exp Biol 38:851–861

Muraleedharan D, Prabhu V (1981) Hormonal influence on feeding and digestion in a plantbug, *Dysdercus cingulatus* and a caterpillar, *Hyblae puera*. Physiol Entomol 6:183–189

Nolen TG, Hoy RR (1984) Initiation of behaviour by single neurons: the role of behavioural context. Science 226:992–994

Remorov VV (1982) Localization of the receptors determining the choice of the food plants by caterpillars (Lepidoptera). Ent Obozr 61:463–471

Retnakaran A, Lauzon H, Fast P (1983) *Bacillus thuringiensis* induced anorexia in the spruce budworm, *Choristoneura fumiferana*. Entomol Exp Appl 34:233–239

Saxena KN, Schoonhoven LM (1978) Induction of orientational and feeding preferences in *Manduca sexta* larvae for an artificial diet containing citral. Entomol Exp Appl 23:72–78

Saxena KN, Schoonhoven LM (1982) Induction of orientational and feeding preferences in *Manduca sexta* larvae for different food sources. Entomol Exp Appl 32:173–180

Schoonhoven LM (1967) Chemoreception of mustard oil glucosides in larvae of *Pieris brassicae*. Proc K Ned Akad Wet C 70:556–568

Schoonhoven LM (1969a) Sensitivity changes in some insect chemoreceptors and their effect on food selection behaviour. Proc K Ned Akad Wet C72:491–498

Schoonhoven LM (1969b) Amino-acid receptor in larvae of *Pieris brassicae* (Lepidoptera). Nature 221:1268

Schoonhoven LM (1972) Plant recognition by lepidopterous larvae. Symp R Entomol Soc Lond 6:87–99

Schoonhoven LM (1976) On the variability of chemosensory information. Symp Biol Hung 16:261–266

Schoonhoven LM (1977) On the individuality of insect feeding behaviour. Proc K Ned Akad Wet C 80:341–350

Schoonhoven LM (1981) Chemical mediators between plants and phytophagous insects. In: Nordlund DA (ed) Semiochemicals: Their Role in Pest Control. John Wiley, New York, pp 31–50

Schoonhoven LM, Dethier VG (1966) Sensory aspects of host-plant discrimination by lepidopterous larvae. Arch Neerl Zool 16:497–530

Schoonhoven LM, Tramper NM, van Drongelen W (1977) Functional diversity in gustatory receptors in some closely related *Yponomeuta* species (Lep.). Neth J Zool 27:287–291

Scott TR, Chang F-CT (1984) the state of gustatory neural coding. Chem Senses 8:297–314

Sieber R, Benz A (1978) The influence of juvenile hormone on the feeding behaviour of last-instar larvae of the codling moth, *Laspeyresia pomonella* (Lep., Tortricidae) Experientia 34:1647–1650

Simmonds MSJ, Blaney WM (1984) Some neurophysiological effects of azadirachtin on lepidopterous larvae and their feeding response. In: Schmutterer H, Ascher KRS (eds) Natural Pesticides from the Neem Tree and Other Tropical Plants. Proc 2nd Int Neem Conference. Rauisch-Holzhausen Castle, Germany, pp 163–179

Simmonds MSJ, Blaney WM, Delle Monache F, Marquina Mac-Quhae M, Marini Bettolo GB (1985) Insect antifeedant properties of anthranoids from the genus *Vismia*. J Chem Ecol 11:1593–1599

Smith DV, Bowdan E, Dethier VG (1984) Information transmission in tarsal sugar receptors of the blowfly. Chem Senses 8:81–101

Städler E (1982) Sensory physiology of insect-plant relationships-round-table discussion. In: Visser JH, Minks AK (eds) Proc 5th Int Symp Insect-Plant Relationships. Pudoc, Wageningen, pp 81–91

Städler E, Hanson FE (1975) Olfactory capabilities of the "gustatory" chemoreceptors of the tobacco hornworm larvae. J Comp Physiol 104:97–102

Städler E, Hanson FE (1976) Influence of induction of host preference on chemoreception of *Manduca sexta:* behavioural and electrophysiological studies. Symp Biol Hung 16:267–273

Thorsteinson AJ (1960) Host selection in phytophagous insects. Annu Rev Entomol 5:193–218

Torii K, Morii K (1948) Studies on the feeding habit of silkworm. Bull Res Inst Seric Sci 2:3–12

van der Molen JN, Veenman CL, Nederstigt LJA (1985) Variability in blowfly taste responses. J Comp Physiol A 157:211–221

van Drongelen W (1979) Contact chemoreception of host plant specific chemicals in larvae of various *Yponomeuta* species (Lepidoptera). J Comp Physiol 134:265–279

van Drongelen W, van Loon JJA (1980) Inheritance of gustatory sensitivity in F_1 progeny of crosses between *Yponomeuta cagnagellus* and *Y. malinellus* (Lepidoptera). Entomol Exp Appl 28:199–203

Waldbauer GP (1962) The growth and reproduction of maxillectomized tobacco hornworms feeding on normally rejected non-solanaceous plants. Entomol Exp Appl 5:147–158

Waldbauer GP (1964) The consumption, digestion and utilization of solanaceous and non-solanaceous plants by larvae of the tobacco hornworm, *Protoparce sexta* (Johan) (Lepidoptera: Sphingidae). Entomol Exp Appl 7:253–269

Waldbauer GP, Fraenkel G (1961) Feeding on normally rejected plants by maxillectomized larvae of the tobacco hornworm, *Protoparce sexta* (Lepidoptera: Sphingidae). Ann Entomol Soc Am 54:477–485

Waldbauer GP, Cohen RW, Friedman S (1984) Self-selection of an optimal nutrient mix from different diets by larvae of the corn earworm, *Heliothis zea* (Boddie). Physiol Zool 57:590–597

Wieczorek H (1976) The glycoside receptor of the larvae of *Mamestra brassicae* L. (Lepidoptera, Noctuidae). J Comp Physiol 106:153–176

Winstanley C, Blaney WM (1978) Chemosensory mechanisms of locusts in relation to feeding. Entomol Exp Appl 24:550–558

Chapter 7

Chemoreception in the Fly: The Search for the Liverwurst Receptor

Frank E. Hanson*

The history of insect sensory biology took an important turn in 1947 when a fly landed on an open-faced liverwurst sandwich. It sampled the substrate and began ovipositing. This was not just any liverwurst sandwich; it was to have been the lunch of one of the world's most astute observers of insect behavior, who did not let this seemingly trivial event go unnoticed. The greater question soon became obvious: why did the fly oviposit on the liverwurst and not on the lettuce, bread, or sill of the open laboratory window, which had also been landed upon? The answer was not immediately at hand, and thus the search for the liverwurst receptor began.

The early studies were simple, but elegant. Flies mounted on applicator sticks were lowered until their tarsi touched an experimental solution. Any sensory organs located on the legs could then sample the chemicals in the substrate in a manner controlled by the experimenter. An acceptable solution is indicated by the extension of the proboscis. A quantitative measure of the acceptability of the chemical is simply the percentage of positive responses. As the proboscis extends, a second set of hairs on the margins of the labellum touch the stimulus and reveal their sensory function. Acceptability to this second set of sense organs is indicated by the spreading of the labellar lobes, whereas retraction of the proboscis shows the solution to be unacceptable. Opening the labellum uncovers yet a third set of chemosensory organs, the oral papillae, which finally elicit drinking if the solution is acceptable. Thus three sets of chemosensory organs are activated sequentially and can be independently tested for responsiveness (see review by Dethier 1976). In addition, there are similar chemoreceptors on the wings (Wolbarsht and Dethier 1958; Angioy et al. 1981) and ovipositor (Wolbarsht and Dethier 1958; Wallis 1962; Rice 1977).

*Department of Biological Sciences, University of Maryland Baltimore County, Catonsville, Maryland 21228, U.S.A.

Structure

The nature of these organs elicited considerable interest. Dethier (1955a) showed that a drop of concentrated sugar solution applied to the tip of a single hair was sufficient to elicit a behavioral response. Light microscopy revealed that the shafts of responsive hairs contain two longitudinal compartments. One chamber is hollow and fluid-filled; the other contains two dendrites that extend from the cell bodies at the base of the hair to the (inferred) pore at the tip. That two functions, acceptance and rejection, could be coded by two sensory cells indicated to the scientific world at that time that one cell must be sensitive to all acceptable compounds and the other to all rejected compounds. This once again demonstrated the efficiency in form and function of the invertebrate nervous system.

Such a rosy-hued picture could not last long. The electron microscope (EM) could see where the light microscope could not; where two dendrites had previously been seen, four appeared in the EM (Larsen 1962; Larsen and Dethier 1965). Most other observations through the light microscope were confirmed and elaborated on. The dendrites are contained inside the thick-walled tube that runs the length of the hair alongside the hollow, fluid-filled cavity. The significance of the presence of two chambers proved to be more than just structural. The two cavities are in contact with the brush-border-lined sinuses of two different cells at the base of the hair. There is evidence that the fluids in the cavities are supplied and regulated by these cells (Phillips and VandeBerg 1976). This is important for the electrical properties of the sensillum, as will be seen later.

The EM also provided the ability to visualize the terminal pore. The first one successfully photographed in some detail was in the stablefly, *Stomoxys;* the pore was seen as a simple 100-nm hole devoid of structural details and located on the side of the hair just below the tip (Adams et al. 1965). The pore in other species turned out to be anything but simple. In *Phormia* and *Calliphora,* the pictures of Stürckow et al. (1967, 1973) indicate that the tip is a bifurcate structure with a pore that may open and close in response to various external stimuli. Other interpretations of similar pictures have also been proffered (Dethier 1976).

With the advent of high-resolution scanning electron microscopes (SEMs), van der Wolk (1984; van der Wolk et al. 1984) was able to detail further the changes in the tip and correlate them with the types and concentrations of stimuli. Her conclusions are that the cuticle comprising the tip is motile and permits the pore to open and close repeatably. When open, the tip bifurcates to form two prongs (Figure 7.1). The pore is in one of the prongs and points slightly laterally. Normally the pore is open in air; stimulation with salts at high concentration or sugars at any concentration tends to keep it open, whereas dilute salts or water will decrease its size or close it. When the pore closes, the second prong disappears as if cuticular material shifts position to collapse the rim of the pore. The degree of closing depends on the type of salt and on the concentration, but does not appear to correlate well with any single pa-

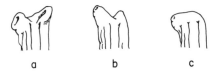

Figure 7.1. One interpretation of the morphological plasticity of the tip of the chemo-sensory hair. Side view of a largest labellar hair of *Calliphora vicina*. Shape of the tip varies with stimulus. a. Air (no stimulus), sugars, and strong salts all tend to open the pore. b. Dilute salts and water partially close it. c. Maximal closure is caused by 96% ethanol. (Courtesy of van der Wolk FM.)

rameter. Thus simple physicochemical mechanisms, such as osmotic pressure changes, are probably ruled out. Other interesting details may also be seen at the tip, summarized by the diagrammatic view presented in Figure 7.2. Inside the pore near its base is a sieve plate guarding the entrance to the dendrite-containing cavity, with a grating small enough (5 nm mesh) to exclude the smallest of pathogens. On the second prong appear small pores (DFLe), in-terpreted as portals through which the dendrite-free lumen may communicate with the exterior. Wax pores (We) can also be seen scattered over the surface. Thus the tip of the hair is far from the simple opening that it was once thought to be, but instead is full of rich detail. Its dynamic motility may be the cause of some of the response variability seen in electrophysiological experiments (de Kramer and van der Molen 1980).

The Four Chemosensory Cells

Four dendrites have been found in all fly contact chemosensory hairs examined. Different species use these cells differently to code for various response mo-dalities. Even among classes or sets of morphologically similar sensilla on a single animal, there are differences in function. However, some response func-tions seem to be common to almost all contact chemosensory organs: sensitivity to sugars, salts, and in many cases, water. The classic blowfly literature has named the four cells based on the functions that were first discovered: sugar, salt, water, and anion (or second salt) cell. Since those early discoveries, other functions have been found for each cell, but yet the classic nomenclature en-dures. Although it is somewhat inaccurate and may be misleading for the no-vitiate, it has communicative value for the cognoscenti and therefore shall be largely retained in that which follows.

The Sugar Cell

The earliest behavioral data of Dethier (reviewed in 1955a, 1976) showed that flies responded positively to many sugars. The sensitivities to the various sugars vary widely: for example, 0.01 *M* fructose or sucrose will elicit threshold re-

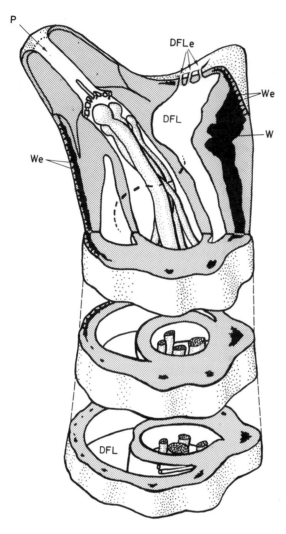

Figure 7.2. Diagrammatic view of the internal features of the tip of the chemosensory hair of *Calliphora vicina*. The hemisected cutaway view (top) shows four dendrites in one compartment, which is bounded by a sieve plate just below the ca. 100 nm diameter pore (P). The larger, dendrite-free lumen (DFL) crosses under the dendrite-containing lumen near the tip and communicates with the exterior by exit ports (DFLe). Longitudinal wax canals (W) vent to the outside via wax exit ports (We). Cross-sections (bottom) represent the structure in the shaft of the hair. (Courtesy of van der Wolk FM.)

sponses in ca. 50% of the flies tested, whereas to obtain this level of response using mannose requires a 5 M solution. In classic receptor fashion, the magnitude of the response is proportional to the logarithm of the concentration over the dynamic range (Figure 7.3). Some sugars, such as fucose, stimulate strongly but are nonnutritive.

The early electrophysiological studies (Hodgson and Roeder 1956; Hodgson 1957) showed that a single cell within the sensillum is active when sugar solutions are applied to the tip. Concentration–response curves showed that action potential frequency is proportional to the logarithm of concentration, just as the behavioral studies predict (see Figure 7.6). The intensity of response to different sugars more or less parallels the behavioral sensitivity hierarchy. Most sugars elicit activity from the same cell. This suggested that there may be a single receptor mechanism that is relatively nonspecific for carbohydrates, or that there are many types of receptor sites (acceptor molecules) on the membrane.

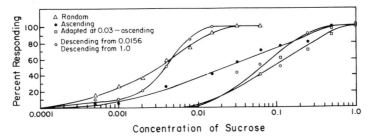

Figure 7.3. Behavioral concentration–response curve to sucrose by *Phormia regina*. The percentage of flies that respond with a proboscis extension when tarsi are dipped into a sucrose solution increases with concentration. Each curve plots the result of a different protocol of stimulus presentation. (After Dethier VG. Biol Bull 103:178–189, 1952; and The Hungry Fly. Harvard University Press, 1976.)

The question of how many acceptor sites are present on the membrane is important biologically because this is a way for animals to increase the breadth of sensitivity within the constraints of a four-cell sense organ. More recent behavioral data suggest that there are at least two classes of sugars. By testing sugars in combinations, two groups were obtained: One group of sugars was shown to synergize when paired with other members of the same group, but to inhibit competitively members of the other group (Dethier 1955a; Dethier et al. 1956). Electrophysiological evidence also agreed with this assessment (Omand and Dethier 1969). These data imply that there are two sites that differentially bind the various sugars and are differentially activated by them. Evidence that the two sites are indeed independent was first presented by Shimada et al. (1974). Several protein modifying reagents block one site but not the other (Figure 7.4). For example, after treatment with *p*-chloromercuribenzoate (PCMB) (a sulfhydryl blocking reagent), a normal response to fructose was

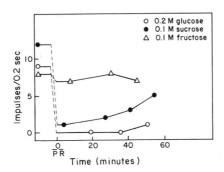

Figure 7.4. Separation of pyranose and furanose sites of the sugar receptor of the fleshfly. Response of the sugar cell to glucose and sucrose diminishes after treatment with 0.5 m*M* *p*-chloromercuribenzoate (PCMB, or *p*-*MB*), whereas the response to fructose is unaffected. P, 5-min treatment with 0.5 m*M* PCMB; *R*, rinse with distilled water. (Reprinted with permission from Shimada I, Shiraishi A, Kijima H, Morita H. Separation of two receptor sites in a single labellar sugar receptor of the flesh-fly by treatment with *p*-chloromercuribenzoate. J Insect Physiol 20 © 1974, Pergamon Press, Ltd.)

elicited whereas the response to glucose decreased by 80%. By systematically studying all the combinations, they concluded there is a "furanose site" responding most strongly to D-fructose, D-fucose, and D-galactose, and a "pyranose site" that would react primarily to D- and L-glucose, D- and L-arabinose, L-fucose, L-sorbose, and D-xylose. The requirements for binding strongly to the pyranose site are equatorial hydroxyl groups at the C-2, C-3, and C-4 positions of the carbon skeleton; different configurations and polar groups at various sites enhance or diminish the stimulating power of the candidate compound (see Dethier 1976 for review). The requirements of the furanose site are not as well understood.

Yet a third site, the "aliphatic carboxylate site," was deduced by Shimada and Isono (1978), based on data from differential depression by various pharmacological reagents. Certain amino acids had previously been shown to elicit behavioral responses and stimulate the sugar cell (Shiraishi and Kuwabara 1970; Goldrich 1973; Shimada 1978); this amino acid sensitivity is partly mediated by this site. The stereospecificity of the furanose site and the aliphatic carboxylate site for amino acids and small peptides has been detailed (Shimada and Tanimura 1981). An accessory site of the latter has been proposed that is specific for the Glu moiety of glutamyl dipeptides (Shimada et al. 1983).

To complicate the sugar receptor even further, a fourth independent site has been proposed to account for the variations in 4-nitrophenyl-α-glucoside sensitivity (Wieczorek 1981; Wieczorek and Köppl 1982).

The actual mechanism of sugar stimulation is still in the hypothesis stage. Classic theory holds that the stimulus molecule and receptor site form a reversible complex. This results in some change in the permeability of the dendritic membrane, presumably by allosteric modification of the shape of the receptor site that opens an ion pore. Ions that are not electrochemically balanced flow through the pores and become redistributed across the membrane. This flow of ions represents the receptor current that depolarizes the cell and produces action potentials in the region of the dendrite near the cell body.

Current carrying ions must be available in the extra-dendritic fluids. If a sensillum is partially depleted of its ions by a long period of immersion in distilled water, then a salt-sucrose solution becomes much more stimulating than the same concentration of sucrose alone (Broyles et al. 1976). Of the ions tested, Na^+ and K^+ are effective, but not Li^+. Consequently, Na^+ and K^+ were proposed as current-carrying ions for the sugar cell.

The source of electrochemical imbalance across the cell membranes in this sensillum may be slightly unusual. In most conventional neurons, this voltage differential, or "driving force," is the diffusion potential of the concentration imbalance of K^+. In the fly sensillum, however, the extra-dendritic ion concentrations must change with each stimulation. To resolve this dilemma, it has been suggested that the source of the potential across the dendritic membrane may be the electrogenic K^+ pump. It is probably located in the highly folded membranes of the sinus formed at the junction of the two sheath cells (tormogen and trichogen cells) at the base of and continuous with the fluid-filled longitudinal compartment (Hansen and Wieczorek 1981). This pump may produce a potential of 100 mV (Thurm 1974). A low resistance pathway through the length of the fluid-filled compartment (which presumably communicates with the dendritic compartment) provides the voltage differential across the dendritic membrane needed as a driving force for the receptor current.

The nature of the sites themselves is a little less mysterious than before. The pH sensitivity suggests that a site has several ionized groups (Shiraishi and Morita, 1969). Salts inhibit the receptor at higher concentrations, presumably by affecting these sites (Shiraishi and Miyachi 1976). Sulfhydryl bonds are obviously important, at least in the pyranose site, since PCMB has such a pronounced effect. Otherwise, the receptor site and membranes are quite resistant to many severe membrane-active reagents, and recover rapidly from any effects that do occur (Shimada 1975).

At one time there was optimism that the receptor site would become completely characterized. As the information about the sugar receptor accumulated, the characteristics of the pyranose site began to resemble those of an α-glucosidase enzyme. To be sure, enzymatic activity is not needed to explain transduction nor binding of the stimulus by the receptor site. However, it was thought that such enzymatic activity could be used as a tracer permitting the receptor protein to be extracted, purified, and characterized using the powerful techniques of enzymology and protein chemistry. Remembering that Dethier (1955b) had discovered glycosidase activity on the tarsi, Hansen (1969) and others extracted, purified, and characterized the source of this enzymatic activity. They found not just one, but 10 such enzymes, and from other parts of the body surface as well. The fraction having the highest specific activity is from the labellar and tarsal hairs, and its properties most closely resemble those of the receptor. But their calculation that the observed activity can be accounted for by only 30 acceptor molecules dampens any hopes that it may be biochemically isolated and characterized. Thus at the moment, the "glycosidase hypothesis" for the pyranose receptor site can neither be verified nor rejected, but remains an intriguing idea.

The Water-Sugar Cell

Behavioral responses to pure water were some of the earliest observed. Flies deprived of water extend mouthparts and drink when their tarsi or labella contact water. This provided the rationale to search for the water fiber as soon as electrophysiological techniques became available (Wolbarsht 1957). Evans and Mellon (1962a) studied this response in more detail and concluded that in the longest labellar hairs of *Phormia regina* water ordinarily activates a single cell, although occasionally a few spikes of another class are also seen.

The response of this cell to water is strongly influenced by low concentrations of salts, thus prompting the suggestion that it could be considered a salt receptor with a negative concentration–response curve. In actuality, such a curve rises slightly when small amounts of salt are added to pure water, peaking at 0.002 M NaCl, then falling to zero (complete inhibition) at about 0.05 M (Figure 7.5)

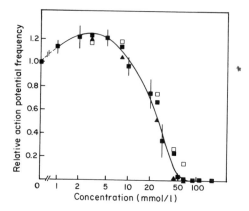

Figure 7.5. Effects of monovalent electrolytes on responses of the water-sugar cell of *Phormia regina*. Responses to each solution are normalized to the responses to pure water. (■), NaCl; (□), LiCl; (▲), KCl. Symbols are mean responses of five cells, except those with standard error bars (± 2 SEM) which are for one experiment using NaCl on seven cells. (After Rees CJC. Proc R Soc Lond Ser B 174:469–490, 1970.)

(Rees 1970). All of the monovalent alkali chlorides act in the same fashion, but the divalent and trivalent cations inhibit at much lower concentrations (0.01 and 0.002 M). Clearly, the cation is the effective agent, as varying the anion does not modify the inhibition curve. Nonelectrolytes also inhibit, but only at much higher concentrations, and roughly in proportion to their osmotic concentration.

The cell is also stimulated by substances other than water. Anesthetics (e.g., halothane) elicit activity in labellar hairs of *Phormia regina* (Dethier, 1972). In some cases, the responses were higher than to water. These responses may provide some insight into the mechanism of function of the water receptor.

Perhaps even more interesting is that the analogous cell in *Phormia terraenovae* responds to sugars (Wieczorek and Köppl 1978). Water responsiveness can be inhibited with low concentrations of salts, and the pure response to sugars can be seen. Extensive analysis showed that only one sugar site is present, and it is nearly identical to the furanose site of the classic sugar cell. Both are primarily stimulated by D-fructose, D-fucose, and D-galactose. Both have similar concentration–response curves (Figure 7.6) with only slightly different Hill plots. The sugar detection capability of this cell is tightly linked to water sensitivity, since those labellar hairs that lack water responsiveness (designated "big marginal" hairs by Wilczek 1967) also do not show this second sugar response (Wieczorek 1980). These hairs only show the response by the classic sugar cell when stimulated by these sugars. Thus the conclusion that the water cell must also be a sugar cell appears inescapable.

The above data are from the labellar hairs of *Phormia regina* and *Phormia terraenovae*. Tarsal hairs are probably different. McCutchan (1969) saw no evidence of a water response in tarsal C and D hairs of *Phormia regina*, although it is possible that a very small spike could have been buried in the baseline noise or inhibited by the electrolyte in her electrode. But in tarsal D hairs of *Calliphora vicina*, both water and sucrose stimulate a cell in classic fashion, and cross-adaptation by the two stimuli is clearly evident (van der Starre 1972). In addition, one or two other cells were reported in both the water and sucrose responses; however, these two cells do not show the concentration–response characteristics of the primary water–sugar cell, nor is cross-adaptation evident, and therefore they are probably not water–sugar receptors.

The molecular mechanism of stimulation by water has evoked much discussion. Because of the size of the water molecule, it is unlikely to be a classic stimulus–acceptor mechanism. Furthermore, PCMB has no effect on the water response, thereby demonstrating that proteins with disulfide linkages are not involved (Shimada et al. 1974). Rees (1970) hypothesized that electrokinetic streaming potentials could occur if water and ions flow through a semipermeable porous membrane with embedded charges. The driving force for the flow would be an osmotic gradient between the dendrite and the stimulus. The calculations of Wieczorek (1980), however, indicate that the potential generated at the tip of the dendrite is not likely to be more than 10 mV. Because of transmission losses down the dendrite, this would not be sufficient by itself to depolarize the membrane at the spike generating region near the base of the hair. It could, of course, contribute to potentials generated by other mechanisms.

In view of the above, a new theory may be needed to explain water detection.

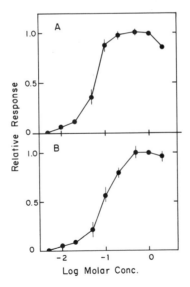

Figure 7.6. Concentration–response curves to D-fructose. A. Water-sugar cell. B. Sugar cell. Data for (A) and (B) were recorded simultaneously from the same sensillum (spikes from the two cells are distinguishable and therefore can be counted independently). Response to each concentration is normalized to the 1 *M* D-fructose response. Points represent the mean ± SEM of 12 labellar sensilla of *Phormia terraenovae*. (After Wieczorek H, Köppl R. J Comp Physiol 126:131–136, 1978.)

Because stimulation by water also affects ion concentrations around the dendritic membrane, perhaps a slight modification of the mechanisms that control the salt cell also apply to the water cell (see below: Salt Cell). Thus the addition of water would presumably be equivalent to removal of ions, with the attendant effects on the charged membrane. But to move from hypothesis to understanding will require many experiments, and the elucidation of the mechanism of water detection will be an interesting story.

The Classic Salt Cell

The first cell to be recorded (Hodgson et al. 1955) and certainly the one from which good recordings are most easily obtained is the "classic salt cell" in the labellar and tarsal hairs of *Phormia regina*. It is the largest spike in the labellar hairs and thus originally termed the "L" spike by Hodgson and Roeder (1956). The high ionic content of the stimulating solutions suppresses the water cell for "clean" records, and provides good electrical conductivity resulting in excellent signal-to-noise ratios. It responds to monovalent cation salt solutions as a classic phasic-tonic sensory cell with textbook concentration–response curves, and thus is an excellent preparation for study or demonstration for students.

Early characterizations of this cell were made by Evans and Mellon (1962b) and Rees (1968). In addition, Gillary (1966b) showed that the alkali halides all stimulate this cell in a similar manner such that complete cross-adaptation is seen among the alkali cations. Nevertheless, each elicits a slightly different concentration–response curve, suggesting that the receptor binding characteristics are different for each. Anions have a strong effect on the response, and thus the commonly used term "cation receptor" is not strictly accurate. The response to salts is remarkably independent of pH over the range 3–11 (Gillary 1966a). This prompted Evans and Mellon (1962b) to suggest that a phosphate group must be involved at the receptor site in order to explain its pH insensitivity. Stimuli having pH values above or below this range, however, elicit generalized activity. The cell's responses are very strongly dependent on temperature (Q_{10} of 3 to 4) such that even evaporation of the stimulating solution from the tip of the electrode cools it enough to affect the results.

Divalent salts affect the cell in a much different manner than monovalent salts (Rees and Hori 1968). When mixed with monovalent cations, divalent cations synergize at low concentrations (from 0 to 0.1 M) and inhibit at higher concentrations (Figure 7.7). In the absence of monovalent ions, salts of divalent cations inhibit any spontaneous activity of the salt cell and a negative potential (presumably a hyperpolarization of the cell) can be recorded. These effects are not exclusively the province of the divalent cations, since choline chloride also hyperpolarizes and inhibits, but at much higher ($\times 5$) concentrations (Rees 1968). In contrast to these responses of the labellar hairs, the oral papilla is strongly activated by $CaCl_2$ (Dethier and Hanson 1965).

As in the labellar hairs described above, the tarsal chemosensory sensilla

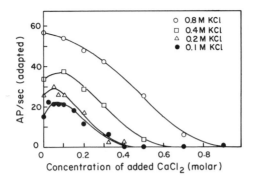

Figure 7.7. Effects of a divalent cation on reponses of the classic salt cell in a labellar hair of *Phormia regina*. Data are mean response frequencies of five cells responding to four different concentrations of KCl when mixed with increasing concentrations of added CaCl$_2$. Frequencies were measured 400 msec after onset of stimulus when response had adapted to steady tonic phase. (Reprinted with permission from Rees CJC, Hori N. The effect of electrolytes of the general formula XCl$_2$ on the response of the type 1 labellar chemoreceptor of the blowfly *Phormia*. J Insect Physiol 14 © 1968, Pergamon Press, Ltd.)

also have two (occasionally three) cells that respond to salts (Figure 7.8), but only one that always shows the classic concentration–response curve (Mc-Cutchan 1969; Hanson, Cearley, and Kogge unpublished). The response is strongly phasic-tonic: Figure 7.9 shows the phasic portion adapting within 100 msec, and changing abruptly to a gently sloping tonic phase. The relationship between the two phases can be seen more clearly on a semilog plot (Figure 7.10) showing that each phase has a different decay rate. This implies that they have different mechanisms. Each portion of the curve is fit quite well (R^2 = 0.83 and 0.94, respectively) by equations of the form $Y = ae^{-bx}$, suggesting

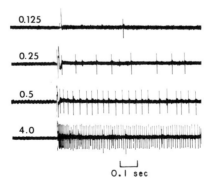

Figure 7.8. Response of a tarsal salt cell of *Phormia regina*. The stimuli are four concentrations of NaCl indicated at left of each trace. The large spikes are responses of the classic salt cell; the smaller spikes are unidentified, but presumably from the second salt cell.

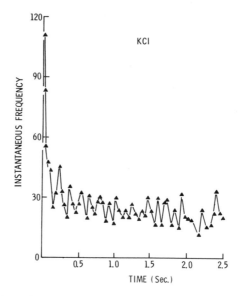

Figure 7.9. Adaptation of a tarsal salt cell (single response). Instantaneous frequency (action potentials per second) of a tarsal sensillum of *Phormia regina* to 0.5 *M* KCl. Data points represent the reciprocal of raw interspike intervals.

exponential decay processes occur in both. The constants *a* and *b* have been calculated for several concentrations of alkali chlorides (Hanson, Cearley, and Kogge unpublished).

Increasing the concentration of the salt elicits a greater response over a certain range. The concentration–response curves for the tarsal D hairs of *Phormia regina* (Figure 7.11) are quite similar to those obtained by Gillary (1966c) from

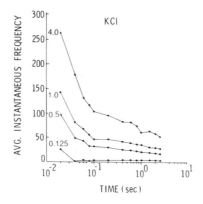

Figure 7.10. Adaptation of the tarsal salt cell (averaged responses), plotted on a logarithmic time axis. Each data point represents the average action potential frequency of ca. 250 trials obtained from the ten D-type sensilla of the right foretarsi of three animals of *Phormia regina*. Stimuli are KCl solutions of molarities indicated at left of each trace.

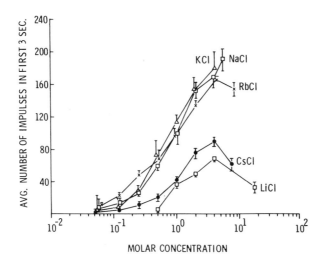

Figure 7.11. Concentration-response curve of the tarsal salt cell. Each point is the average of ca. 180 trials (± SEM) from the ten D-type sensilla of the right foretarsi of eight animals of *Phormia regina*.

the labellar hairs. The response hierarchy is the irregular series $K^+ = Na^+ = Rb^+ > Cs^+ > Li^+$ in both types of sensilla. Because of the close similarity, it is tempting to speculate that all of the fly's salt cells are the same. However, this is probably not the case, since McCutchan (1969) reported that other sizes of hairs on the tarsi of *Phormia regina* have slightly different concentration–response curves. Other workers (e.g., Maes and den Otter 1976; Crnjar 1981) have also shown that different size classes of labellar hairs differ in their responses to salt.

Different species, even closely related ones, have differences in the functional properties of their salt cells. For example, den Otter (1972a) found a different response hierarchy in *Calliphora;* the most common one is: $K^+ > Rb^+ > Cs^+ > Na^+ > Li^+$. But he also reported that different hairs have different response hierarchies, which is a result that would be lost in the normal approach in which responses are averaged. This could explain the behavioral data that flies can discriminate among the different salts (Maes and Bijpost 1979; Busse and Barth 1985).

Theories of the mechanism of taste stimulation by salt start from the classic theory of stimulus–receptor interaction. The stimulus, S, combines reversibly with the specific, unoccupied receptor site, R, to form a complex, RS:

$$S + R \rightleftarrows RS$$

The receptor response would then be proportional to the number of filled sites, RS (Evans and Mellon 1962b), although Heck and Erickson (1973) suggest that instead it may be proportional to the rate at which the forward reaction occurs.

Using this concept as a basis, Evans and Mellon (1962b) proposed that the salt receptor data fit the Beidler (1954) adsorption theory of taste stimulation.

They obtained an equilibrium constant, K, from the tonic phase of the response. The relative free energy change, ΔG, can then be derived from the equilibrium constant. The ΔG values calculated were between -0.06 and -0.7 kcal mole^{-1}, values that indicate weak physical binding rather than chemical bonds. However, difficulties arise when attempts are made to show that the data of the salt receptor actually fit the theory (Gillary 1966a). Small variations in response measurements result in large variations in K. Thus the applicability of the adsorption theory to the mechanism of the blowfly salt receptor function is presently indeterminate.

The classic receptor theory also predicts that the concentration–response curve will be S-shaped (e.g., Figure 7.6) with a response maximum (R_m) at which the receptor sites become saturated, which in turn depends on the number of receptor sites, R. Further increases in stimulus concentration should not further increase the response, and unless desensitization occurs, the response level should plateau at R_m. The concentration–response curves of the blowfly tarsal salt receptor (Figure 7.11) are different than predicted from the classic theory in that no plateau is seen, and two of the five salts (Cs^+ and Li^+) show a peak and then a decrease in activity at supraoptimal concentrations.

Stimuli of many different types activate this receptor (Dethier 1980) making it very unlikely (1) that a single site could be so universally accepting, or (2) that there could be so many specific sites on this neuron. For this reason plus the differences between predicted and observed concentration–response curves, alternatives to the classic theory have been sought that include mechanisms of activation that are nonspecific.

One alternative theory has been put forward by den Otter (1972b, c). This hypothesis holds that the dendritic membrane is comprised of amphipathic (long hydrophobic chain with hydrophilic head) molecules with the hydrocarbon chains extending into the membrane. The hydrophilic heads form a continuous, densely packed surface. The negatively charged heads would be cross-linked by divalent cations for membrane stability. Because of thermal agitation, random breaks would be continually forming and resealing; these would be temporary "statistical pores," which are to be distinguished from "organized pores" such as sodium channels in classic nerve membranes. These statistical pores would serve as channels for current carrying ions and would likely be nonspecific for ion species. Monovalent cations introduced as a stimulus would increase the frequency and size of these statistical pores by neutralizing charges on the hydrophilic heads and displacing the divalent ions. As a result, cross-linking would decrease and there would be less compact packing of the membrane molecules. Introduced anions would contribute by neutralizing charges on the divalent cations. On the other hand, externally added divalent cations would tend to stabilize the membrane, increasing the membrane resistance when supplied alone, or counteracting the effects of monovalent ions when added in mixtures.

The above hypothesis is based on physical chemical experiments on electrolytic biocolloids (references in den Otter 1972a–c) showing that Ca^{2+} binds phosphate biocolloids. If another salt is also added, the Ca^{2+} binding is counteracted at low concentrations of the second salt but synergized at high concentrations.

Den Otter proposed that the salt receptor membrane is functionally analogous to the biocolloid model membrane, and the destabilization by monovalent salts would open statistical pores. The concentration–response curve of the added monovalent cation would ascend, peak, and descend as the concentration increases (Figure 7.12). This family of curves may describe responses of different cells to increased concentrations of salts. Over the range C_1 to C_2 in Figure 7.12, the cell that is positively correlated with concentration (salt cell) is depicted as curve 1 or 2, whereas the cell that is negatively correlated with salt concentration (such as a water cell) is depicted as curve 4 or 5. Alternatively, the figure could represent six different salts applied to the same cell. Note that these hypothetical curves resemble the actual concentration–response curves in Figures 7.5, 7.7, and 7.11 over the solubility ranges of the salts.

This model appears to be consistent with much of the extant data, such as the concentration–response curves, cation stimulation, anion effects, and inhibition by divalent cations. In addition, the same mechanism could explain

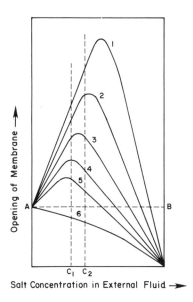

Figure 7.12. Theoretical curves proposed by the "statistical pore" theory. Curves represent the degree of opening of the membrane with changes in salt concentration. Point A represents the intermediate level of packing of membrane molecules in unstimulated cells. Above the level AB, membranes become less compact and more "statistical pores" form, thus increasing permeability. Below the level AB, the membrane becomes more compact and permeability decreases. The curves may illustrate the degree of opening of the membrane of an individual taste cell resulting from stimulation by increasing concentrations of 5 different alkali salts (salts 1 to 5). Alternatively, it could represent the degree of opening of membranes of 5 different types of salt cells (cells 1 to 5) to increasing concentrations of the same salt. Curve 6 represents a membrane becoming more compact when treated with alkaline earth salts of quinine hydrochloride. (Reprinted with permission from den Otter CJ. Interactions between ions and receptor membrane in insect taste cells. J Insect physiol 18 © 1972, Pergamon Press, Ltd.)

the response of water cells if the effect of water is to dilute the ion concentration found naturally in the tip chamber. The water cell would have its activity range at concentrations below that found normally in the tip chamber fluid, as in Figure 7.12, curve 4 or 5.

The "statistical pore" model only attempts to explain the opening of the pores permitting passage of current carrying ions. It assumes the presence of an electrochemical potential to drive the current carrying ions through the pores to depolarize the cell.

An alternative theory that does include an explanation of the receptor potential is the "modification of surface potential" model of Kurihara and colleagues (Miyake and Kurihara 1983; see references in Yoshii and Kurihara 1983). The model in its simplest form is diagrammed in Figure 7.13. The resting membrane potential (E_M) is the difference in potential between the outer bulk phase and the inner (cytoplasmic) bulk phase, separated by the dendritic membrane (Figure 7.13A). However, this is an oversimplified view that does not account for the embedded charges on the inside and outside surfaces of the membrane. These charges could be structural components of the membrane, e.g., phosphate groups on hydrophilic heads of the membrane lipids. These charges produce a negative field (ψ) at the surface layer that extends into each bulk phase, decreasing exponentially with distance (Figure 7.13B). On stimulation by salts

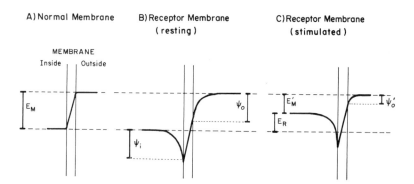

Figure 7.13. Representation of the surface potential theory. Schematic diagrams illustrate the theoretical potential fields in the proximity of two types of cell membrane. (A) Normal membrane with ion channels: the potential inside the cell (lower dashed line) is more negative than that outside (upper dashed line), with, ideally, a linear gradation through the membrane. (B) Recepter membrane, or membrane without ion channels: negative charges embedded on inner and outer surfaces result in surface potentials (ψ) that extend into each bulk phase, decreasing exponentially with distance. (C) Salts in stimulating solution adsorb onto receptor membrane and partially neutralize the external embedded surface charges. This positive-going change in external surface potential raises the potential internally. The change in membrane potential would act as a receptor potential (E_R). E_M, resting membrane potential; E'_M, membrane potential after stimulation; ψ, surface potential on the inside (ψ_i) and outside (ψ_o) of the resting membrane; ψ'_o, surface potential outside after stimulation by salts. (After Yoshii K, Kurihara K. Brain Res 279:185–191, 1983.)

(Figure 7.13C), cations introduced into the external bulk phase would partially neutralize the surface charge on the outside layer of the membrane, thus reducing the outside surface potential (ψ'_o). The difference ($\psi_o - \psi'_o$) would be the receptor potential (E_R), which would raise the potential of the internal bulk phase to a new level (E_M') and partially depolarize the dendrite.

The stimulating ions would be much more effective at screening the embedded charges if they were adsorbed directly onto (or into) the membrane and therefore in close proximity to the charges to be neutralized. Different cells having different types of membranes could adsorb certain ions preferentially; this would be the basis for specific sensitivities. The overall effect is that a cation which is preferentially bound to the outside of the membrane provides a ΔE_M, or receptor potential (E_R in Figure 7.13C).

The basis of the activity of the water cell has also been explained by this model (Yoshii and Kurihara 1983). When the sensillum is not being stimulated, the membrane of the water cell must adsorb cations found normally in the dendritic chamber. Hence the membrane outer layer acquires a positive surface charge. The effects of this surface charge are suppressed by the ionic strength (screening effect) of the normal extracellular salts. Stimulating the sensillum with water presumably dilutes the ions in the extracellular medium, thus decreasing the screening effect. This would result in a positive displacement of the overall membrane potential and provide a receptor potential to stimulate the cell. Any cations in the applied stimulus would contribute to the screening effect, thus accounting for inhibition of the water response by cations (see Water-Sugar Cell, Figure 7.5).

The above theories have been proposed because the classic theory falls short of explaining salt and water reception. For the sugar cell, which is stimulated by comparatively large molecules rich in three-dimensional structure, the classic receptor binding model appears to be a reasonable explanation. It is more difficult to imagine the same for small molecules such as water or monovalent cations, particularly when a broad array of very different compounds also activates these cells (Dethier 1972, 1980).

The surface potential theory was proposed to explain vertebrate chemosensory data, and may need some theoretical and experimental work to shape it for insects. Amakawa (1978) tested its application to the blowfly sugar receptor, but the resulting data did not fit the hypothesis. However, the theory seems much more applicable to the salt and water cells, although specific experimental verification is needed.

The Second Salt Cell

A second spike is often seen in response to salt. The frequency of this spike is not usually correlated with salt concentration (den Otter 1972a), and consequently most workers ignore it. Steinhardt (1965) suggested that it might be a salt cell primarily responsive to anions (thus the origin of the term "anion cell"), but that is inhibited by many kinds of cations and therefore rarely seen.

Dethier and Hanson (1968) noted that it becomes active when sodium salts of certain fatty acids are used as stimuli. These authors simply referred to it as the "fifth cell." Other than these studies, very little attention has been paid this cell. This is unfortunate, because this lack of information handicaps our attempts to view the entire picture of the control of feeding behavior of the fly on the basis of its chemosensory input.

Chemosensory Information and Behavior

The next higher level of understanding of the physiological basis of feeding behavior requires one to look beyond the myriad functional details of the chemosensory cells just reviewed and consider only their outputs as information on which feeding decisions are made. The primary questions are what is the relevant information, and how is it used by the animal?

The answer for the sugar cell and the water–sugar cell seems to be fairly obvious: the observed activity in these cells is highly correlated with behavioral acceptance in hungry or thirsty flies. For the salt cell the answer is not so obvious. At one time it was thought that rejection of a candidate food is mediated by activity in the salt cell, since the behavioral experiments showed that increasing the salt concentration of a salt–sugar mixture resulted in a decrease in acceptance. It was then hypothesized that the behavioral decision was based on the ratios of the inputs of the sugar cell (acceptance) and salt cell (deterrence). However, the experimental foundation of this attractive hypothesis began to crumble when Shiraishi and Miyachi (1976) showed that in these experiments the decrease in behavioral acceptance could be accounted for by peripheral inhibition of the sugar fiber by the added salt. The salt cell input, although demonstrably present in great abundance, must have been ignored by the central nervous system of the fly in this experimental situation. One concludes that this cell is not so potent a deterrent cell as once thought.

Nevertheless, the data pointing to the salt cell as a deterrent cell cannot be ignored. Salt stimulation clearly contributes to a central inhibitory state (Dethier et al. 1965, 1968). Also, Dethier (1980) reported that most of the 59 deterrent plant saps and 24 deterrent chemicals that were applied to the sensillum activated the salt cell only. Furthermore, a direct link between rejection and activation of the salt receptor can be shown in certain situations. For example, a normal fly with its proboscis partially extended will retract it when an individual chemoreceptive hair is stimulated with a high concentration of salt ($\geqslant 0.5\ M$), or will fail to extend it if it is already in the resting position (Dethier 1968). On the other hand, intermediate concentrations (e.g., $0.2\ M$) of monovalent salts elicit proboscis extension. Electrophysiological records demonstrate that in both cases 90% or more of the activity is from the salt cell. Hence this cell must make primary contributions to both acceptance and rejection, perhaps depending on its firing rate. Furthermore, behavioral studies have shown that in some nutritional states there is a distinct preference for "salty" stimuli: protein-deprived female *Phormia regina* prefer salt–sugar mixtures to sugar alone. Con-

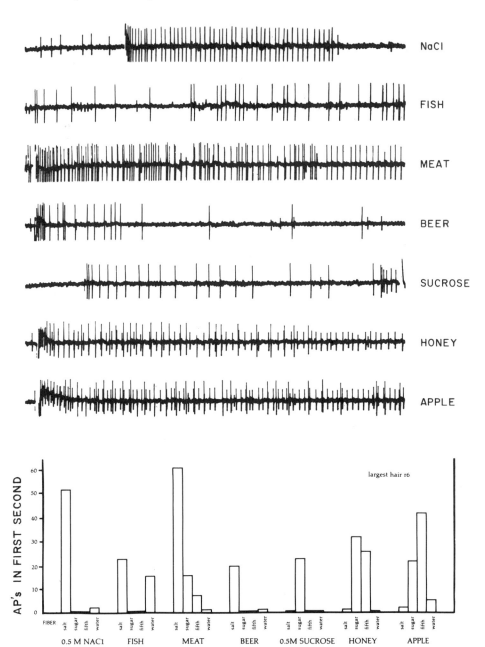

Figure 7.14. Top. Responses to complex stimuli by a D-type (largest) labellar hair of *Phormia regina*. Large spikes in traces 1, 2, and 4 were retouched. Bottom. Graph summarizing the above. Bars represent the number of impulses generated by the salt cell, sugar cell, "fifth" (second salt) cell, and water (water-sugar) cell. (After Dethier VG. J Insect Physiol 20:1859–1869, 1974; and The Hungry Fly, Harvard University Press, 1976.)

sidering that there is no chemosensory cell that responds to proteins per se, perhaps wild female blowflies find protein by selecting food that stimulates both salt and sugar fibers (Rachman 1979; Rachman et al. 1982).

These results show that the early ideas of the sensory code, namely that specific cells only code for either acceptance or rejection, are no longer sufficient (for further discussion, see Chapter 7 of Dethier 1976). Responses to complex mixtures clearly show that the insect obtains information across multiple input channels and integrates them nonlinearly. For example, Dethier (1974) showed that complex substances, such as beer, honey, apple, etc. (all of which are acceptable) elicit complex responses from three or four cells (Figure 7.14). Clearly, many different input patterns can signal acceptability.

The interpretation of these complex codes is one of the challenges facing insect neurophysiology in the future. Smith et al. (1983a) have shown how much information the chemosensory neurons carry. Needed are more sensory coding studies, in which reliable sensory information is correlated with precise behavioral data. This is particularly important for responses to complex stimuli that the fly may encounter in its natural environment. Accordingly, studies of the type initiated by Dethier (1974, 1980) must be extended. Because of the complex nature of the sensory responses to such complex stimuli, computer analysis of the electrophysiological data will become important (Hanson et al. 1986; Frazier and Hanson in press). The multidimensional analysis approach (e.g., see Smith et al. 1983b) may provide yet another avenue for understanding the physiological bases of these behaviors. Perhaps it is within these provinces that we should continue the search for the elusive liverwurst receptor.

References

Adams JR, Holbert PE, Forgash AJ (1965) Electronmicroscopy of the contact chemoreceptors of the stable fly, *Stomoxys calcitrans* (Diptera: Muscidae). Ann Entomol Soc Am 58:909–917

Amakawa T (1978) Effects of cations on the sugar receptor site of the blowfly, *Phormia*. Chem Senses Flavor 3:413–422

Angioy AM, Liscia A, Pietra P (1981) Some functional aspects of the wing chemosensilla in *Phormia regina* (Meig.) (Diptera Calliphoridae). Monit Zool Ital 15:221–228

Beidler LM (1954) A theory of taste stimulation. J Gen Physiol 38:133–139

Broyles JL, Hanson FE, Shapiro AM (1976) Ion dependence of the tarsal sugar receptor of the blowfly *Phormia regina*. J Insect Physiol 22:1587–1600

Busse FK, Barth RH (1985) Physiology and feeding preference patterns of female black blowflies (*Phormia regina* Meigen): modification in responsiveness to salts subsequent to salt feeding. J Insect Physiol 31:23–26

Crnjar R (1981) The chemosensory function of the "bristle" labellar taste sensilla in *Phormia regina* (Meig.) (Diptera Calliphoridae). Monit Zool Ital 15:95–105

de Kramer JJ, van der Molen JN (1980) The pore mechanism of the contact chemoreceptors of the blowfly, *Calliphora vicina*. In: van der Starre H (ed) Olfaction and Taste VII. Information Retrieval, London, pp 61–64

den Otter CJ (1972a) Differential sensitivity of insect chemoreceptors to alkali cations. J Insect Physiol 18:109–131 den Otter CJ (1972b) Interactions between ions and receptor membrane in insect taste cells. J Insect Physiol 18:389–402

den Otter CJ (1972c) Mechanism of stimulation of insect taste cells by organic substances. J Insect Physiol 18:615–625

Dethier VG (1955a) The physiology and histology of the contact chemoreceptors of the blowfly. Q Rev Biol 30:348–371

Dethier VG (1955b) Mode of action of sugar-baited fly traps. J Econ Entomol 48:235–239

Dethier VG (1968) Chemosensory input and taste discrimination in the blowfly. Science 161:389–391

Dethier VG (1972) Sensitivity of the contact chemoreceptors of the blowfly to vapors. Proc Natl Acad Sci USA 69:2189–2192

Dethier VG (1974) The specificity of the labellar chemoreceptors of the blowfly and the response to natural foods. J Insect Physiol 20:1859–1869

Dethier VG (1976) The Hungry Fly. Harvard University Press, Cambridge, MA

Dethier VG (1980) Evolution of receptor sensitivity to secondary plant substances with special reference to deterrents. Am Natur 115:45–66

Dethier VG, Hanson FE (1965) Taste papillae of the blowfly. J Cell Comp Physiol 65:93–100

Dethier VG, Hanson FE (1968) Electrophysiological responses of the chemoreceptors of the blowfly to sodium salts of fatty acids. Proc Natl Acad Sci USA 60:1296–1303

Dethier VG, Evans DR, Rhoades MV (1956) Some factors controlling the ingestion of carbohydrates by the blowfly. Biol Bull Woods Hole Mass 111:204–222

Dethier VG, Solomon RL, Turner LH (1965) Sensory input and central excitation and inhibition in the blowfly. J Comp Physiol Psychol 60:303–313

Dethier VG, Solomon RL, Turner LH (1968) Central inhibition in the blowfly. J Comp Physiol Psychol 63:144–150

Evans DR, Mellon DeF (1962a) Electrophysiological studies of a water receptor associated with the taste sensilla of the blowfly. J Gen Physiol 45:487–500

Evans DR, Mellon DeF (1962b) Stimulation of a primary taste receptor by salt. J Gen Physiol 45:651–661

Frazier J, Hanson FE (in press) Electrophysiological recording and analysis of insect chemosensory responses. In: Miller TA, Miller J (eds) Insect Plant Interactions. Springer-Verlag, New York

Gillary HL (1966a) Stimulation of the salt receptor or of the blowfly. I. NaCl. J Gen Physiol 50:337–350

Gillary HL (1966b) Stimulation of the salt receptor of the blowfly. II. Temperature. J Gen Physiol 50:351–357

Gillary HL (1966c) Stimulation of the salt receptor of the blowfly. III. The alkali halides. J Gen Physiol 50:359–368

Goldrich (Rachman) NJ (1973) Behavioral responses of *Phormia regina* (Meigen) to labellar stimulation with amino acids. J Gen Physiol 61:74–88

Hansen K (1969) The mechanism of insect sugar reception, a biochemical investigation. In: Pfaffmann C (ed) Olfaction and taste III. Rockefeller University Press, New York, pp 382–391

Hansen K, Wieczorek H (1981) Biochemical aspects of sugar reception in insects. In: Cagan RH, Kare MR (eds) Biochemistry of Taste and Olfaction. Academic Press, New York, pp 139–162

Hanson FE, Kogge S, Cearley C (1986) Computer analysis of chemosensory signals.

In: Payne TL, Birch MC, Kennedy CEJ (eds) Mechanisms in Insect Olfaction. Oxford University Press, Oxford, pp 269–278

Heck GL, Erickson RP (1973) A rate theory of gustatory stimulation. Behav Biol 8:687–712

Hodgson ES (1957) Electrophysiological studies of arthropod chemoreception, II. Responses of labellar chemoreceptors of the blowfly to stimulation by carbohydrates. J Insect Physiol 1:240–247

Hodgson ES, Roeder KD (1956) Electrophysiological studies of arthropod chemoreception. I. General properties of the labellar chemoreceptors of Diptera. J Cell Comp Physiol 48:51–76

Hodgson ES, Lettvin JY, Roeder KD (1955) Physiology of a primary chemoreceptor unit. Science 122:417–418

Larsen JR (1962) The fine structure of the labellar chemosensory hairs of the blowfly, *Phormia regina* Meigen. J Insect Physiol 8:683–691

Larsen J, Dethier VG (1965) The fine structure of the labellar antennal chemoreceptors of the blowfly, *Phormia regina*. Proc 16th Int Congr Zool Washington 3:81–83

Maes FW, Bijpost SCA (1979) Classical conditioning reveals discrimination of salt taste quality in the blowfly, *Calliphora vicina*. J Comp Physiol 133:53–62

Maes FW, den Otter CJ (1976) Relationship between taste cell responses and arrangement of labellar taste setae in the blowfly *Calliphora vicina*. J Insect Physiol 22:377–384

McCutchan MC (1969) Responses of tarsal chemoreceptive hairs of the blowfly, *Phormia regina*. J Insect Physiol 15:2059–2068

Miyake M, Kurihara K (1983) Resting potential of the mouse neuroblastoma cells. II. Significant contribution of the surface potential to the resting potential of the cells under physiological conditions. Biochim Biophys Acta 762:256–264

Omand E, Dethier VG (1969) An electrophysiological analysis of the action of carbohydrates on the sugar receptor of the blowfly. Proc Natl Acad Sci USA 62:136–143

Phillips CE, VandeBerg JS (1976) Directional flow of sensillum liquor in blowfly *(Phormia regina)* labellar chemoreceptors. J Insect Physiol 22:425–429

Rachman NJ (1979) The sensitivity of the labellar sugar receptors of *Phormia regina* in relation to feeding. J Insect Physiol 25:733–739

Rachman NJ, Busse FK, Barth RH (1982) Physiology of feeding preference patterns of female back blowflies (*Phormia regina* Meigen): alterations in responsiveness to salts. J Insect Physiol 28:625–630

Rees CJC (1968) The effect of aqueous solutions of some 1:1 electrolytes on the electrical response of the type 1 ("salt") chemoreceptor cell in the labella of *Phormia*. J Insect Physiol 14:1331–1364

Rees CJC (1970) The primary process of reception in the type 3 ("water") receptor cell of the fly, *Phormia terraenovae*. Proc R Soc Lond Ser B 174:469–490

Rees CJC, Hori N (1968) The effect of electrolytes of the general formula XCl_2 on the response of the type 1 labellar chemoreceptor of the blowfly *Phormia*. J Insect Physiol 14:1499–1513

Rice MJ (1977) Blowfly ovipositor receptor neurone sensitive to monovalent cation concentration. Nature 268:747–749

Shimada I (1975) Chemical treatments of the labellar sugar receptor of the fleshfly. J Insect Physiol 21:1565–1574

Shimada I (1978) The stimulating effect of fatty acids and amino acid derivatives on the labellar sugar receptor of the fleshfly. J Gen Physiol 71:19–36

Shimada I, Isono K (1978) The specific receptor site for aliphatic carboxylate anion in the labellar sugar receptor of the fleshfly. J Insect Physiol 24:807–811

Shimada I, Tanimura T (1981) Stereospecificity of multiple receptor sites in a labellar sugar receptor of the fleshfly for amino acids and small peptides. J Gen Physiol 77:23–29

Shimada I, Shiraishi A, Kijima H, Morita H (1974) Separation of two receptor sites in a single labellar sugar receptor of the flesh-fly by treatment with p-chloromercuribenzoate. J Insect Physiol 20:605–621

Shimada I, Maki Y, Sugiyama H (1983) Structure-taste relationship of glutamyl valine, the "sweet" peptide for the fleshfly: the specific accessory site for the glutamyl moiety in the sugar receptor. J Insect Physiol 29:255–258

Shiraishi A, Kuwabara A (1970) The effects of amino acids on the labellar hair chemosensory cells of the fly. J Gen Physiol 56:768–782

Shiraishi A, Miyachi N (1976) The peripheral inhibition of the tarsal sugar receptor by sodium chloride in the proboscis extension response of the blowfly, *Phormia regina*. J Comp Physiol 110:97–110

Shiraishi A, Morita H (1969) The effects of pH on the labellar sugar receptor of the fleshfly. J Gen Physiol 53:450–470

Smith DV, Bowdan E, Dethier VG (1983a) Information transmission in tarsal sugar receptors of the blowfly. Chem Senses 8:81–102

Smith DV, Van Buskirk RL, Travers JB, Bieber SL (1983b) Coding of taste stimuli by hamster brain stem neurons. J Neurophysiol 50:541–558

Steinhardt RA (1965) Cation and anion stimulation of electrolyte receptors of the blowfly, *Phormia regina*. Am Zool 5:651

Stürckow B, Holbert PE, Adams JR (1967) Fine structure of the tip of chemosensitive hairs in two blowflies and the stable fly. Experientia 23:780–782

Stürckow B, Holbert PE, Adams JR, Anstead RJ (1973) Fine structure of the tip of the labellar taste hair of the blowflies, *Phormia regina* and *Calliphora vicina* (Diptera, Calliphoridae). Z Morphol Tiere 75:87–109

Thurm U (1974) Basics of the generation of receptor potentials in epidermal mechanoreceptors of insects. In: Schwartzkopff J (ed) Symposium Mechanoreception. Abh Rhein Westf Akad Wiss 53:355–385

van der Starre H (1972) Tarsal taste discrimination in the blowfly, *Calliphora vicina*. Neth J Zool 22:277–282

van der Wolk FM (1984) The structure of chemosensory sensilla of the fly. Doctoral Dissertation, Leiden University, Netherlands

van der Wolk FM, Koerten HK, van der Starre H (1984) The external morphology of contact chemoreceptive hairs of flies and the motility of the tips of these hairs. J Morphol 180:37–54

Wallis DI (1962) The sense organs on the ovipositor of the blowfly, *Phormia regina*. J Insect Physiol 8:453–467

Wieczorek H (1980) Sugar reception by an insect water receptor. J Comp Physiol 138:167–172

Wieczorek H (1981) Sugar receptors in the labellar taste hairs of the fly. Adv Physiol Sci 23:481–493

Wieczorek H, Köppl R (1978) Effect of sugars on the labellar water receptor of the fly. J Comp Physiol 126:131–136

Wieczorek H, Köppl R (1982) Reaction spectra of sugar receptors in different taste hairs of the fly. J Comp Physiol A: 149:207–213

Wilczek M (1967) The distribution and neuroanatomy of the labellar sense organ of the
    blowfly *Phormia regina*. J Morphol 122:175–201
Wolbarsht ML (1957) Water taste in *Phormia*. Science 125:1248
Wolbarsht ML, Dethier VG (1958) Electrical activity in the chemoreceptors of the
    blowfly. I. Responses to chemical and mechanical stimulation. J Gen Physiol 42:393–
    412
Yoshii K, Kurihara K (1983) Mechanism of the water response in carp gustatory re-
    ceptors: independent generation of the water response from the salt response. Brain
    Res 279:185–191

Chapter 8
The Strange Fate of Pyrrolizidine Alkaloids

DIETRICH SCHNEIDER*

A fairy tale: Once upon a time there was a nice little pyrrolizidine alkaloid molecule. It finds its protective task in the plant rewarding but somewhat dull in spite of the dramatic effects it has upon the big herbivores and the human pharmacophages. Somehow its use after the transfer into the insect is more of an adventure, and particularly fascinating is its role in the social and love life of the Lepidoptera. But the promotion of the growth of a gigantic organ that serves to bring all the moths together really is the crowning role in the life of any molecule. And, after all this is done, the little molecule happily agrees to its disintegration to await a later reassembly and reincarnation, perhaps as a human pheromone molecule?

Plants, as well as other organisms, are composed of chemical substances. In 1891, the pioneer of cytochemistry, A. Kossel, subdivided plant components into primary and secondary ones. Mothes (1980, 1984) quotes from an 1896 lecture of Kossel, who addressed the Berlin Physiological Society (in liberal translation from the German):

> The search and description of those atomic complexes, which are the essence of life are the foundation for the investigation of the life processes. I propose to call the essential components of the cell PRIMARY and those that are not found in all the cells that have the capacity to develop, SECONDARY. The decision whether a substance is a primary or a secondary one is in some cases difficult.

Until our time, many investigators thought that the secondary substances were waste products (these people were named the "waste-product lobby" by Swain 1976), but the majority in the respective disciplines now agree that the secondary substances are "the products of special synthetic activities of specialized cells," are "ecologically important for the producers," "defend them from animals and micro-organisms," and "serve as intra- and interspecific signals" (Teuscher 1984).

---

*Max Planck Institut für Verhaltensphysiologie, D-8131 Seewiesen, F.R.G.

In my chapter I will tell you the fascinating story of one alkaloidal group of such substances. In particular, I will describe the plant-dependent and always risky life of some insects, their pheromones, and the puzzling case of the morphogenesis of a glandular organ in a moth that is controlled by one such alkaloid.

Only briefly will I note the role these alkaloids play in human health and in that of some domestic animals. After all, many secondary plant substances have been since early times—and to some extent still are—elements of medicinal therapy. Such use was often connected with the stimulating or even hallucinogenic power of these substances, of which coffee, opium, and digitalis are only three of many possible examples. Pyrrolizidine alkaloids (PAs) are produced by a variety of plants, in some cases up to remarkable amounts of several percent of the dry weight of some organs (flower heads, seeds). There are some plant families with many PA-containing species, such as the Asteraceae with *Senecio* and *Eupatorium*, the Boraginaceae, and the Fabaceae. Chemically all PAs are either mono- or diesters with a heterocyclic, N-containing, necine-base moiety and a necine-acid moiety (typical examples in Figure 8.1).

Before the 18th and 19th centuries it was not suspected that plants such as *Senecio, Heliotropium* (Boraginaceae), and *Crotalaria* (Fabaceae), if eaten by domestic animals, made them sick, in particular inducing liver diseases, sometimes after a very long incubation time (Bull et al. 1968). Often such a pyrrolizidine alkaloidosis was named after the plant concerned, e.g., senecionosis, but in the course of the last decades a firmer understanding of PA toxicology was established. But the relationships between man and PAs have other facets too. In some parts of the world, such as South Africa, Jamaica, and Mexico,

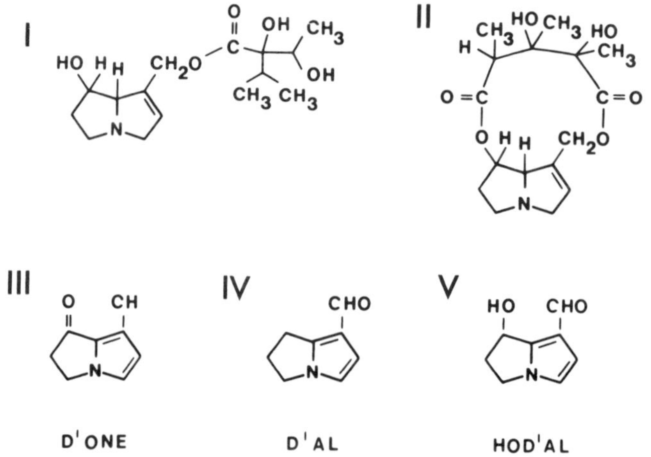

Figure 8.1. Two types of pyrrolizidine alkaloids; (I) monoester (e.g., lycopsamine or heliotrine) and (II) diester (e.g., monocrotaline or seneciophylline), are representatives of this group of secondary plant substances. The following compounds not only render plants containing them more or less unpalatable for herbivores, but are used by some Lepidoptera as precursors for their pheromones: (III) danaidone (D'one); (IV) danaidal (D'al); (V) hydroxydanaidal (HOD'al).

herbal teas are prepared from *Senecio* for their medicinal and hallucinogenic effects. Even in some European countries, PA-containing plant extracts are also still available for medical treatment, such as *Senecio* extracts for their "antidiabetic" or "hemostyptic" powers. Extracts of comfrey (*Symphytum peregrinum,* Boraginaceae) (Radix Consolidae) are sold for medical use, although the dangerous effect of the PA in the extract became well known recently. This toxicity by far offsets any expected benign effects (for literature see Bull et al. 1968; Huxtable 1980; Lüthy et al. 1983; Roitman 1983; Röder 1984).

The noxious effect of the PAs is not a simple one since most of them (but not all!) are eventually hepatotoxic, teratogenic, mutagenic, carcinogenic, and—like all alkaloids—bitter. It became known that the toxicity of the PAs develops only after a metabolic hydrolysis of the ester and desaturation of the necine base, a process that normally occurs first in the intestine and then in the liver of the vertebrates through the action of microsomal enzymes (Figure 8.2). The dehydropyrrolizidines are highly reactive compounds and readily cross-link with purine and pyrimidine bases of DNA or RNA (see citations above). Finally, with respect to the bitter taste of the PAs it is worth remembering that "virtually all compounds of biological origin which are toxic are also bitter tasting to humans" (Brower 1984), although the deadly *Amanita* toadstool is said to have no bad taste.

The question of the "biological meaning" of secondary plant substances, as metabolic by-products, for example, or as substances that are useful in increasing the fitness of the plant, has been violently debated. Over the last two decades, biologists developed the opinion that one of several, or even the main "raison d'être" of secondary plant substances is their antifeedant effect on herbivores (Fraenkel 1959; see also the review by Bernays 1983). Often this organismic interaction has been called co-evolution (Ehrlich and Raven 1967). This has generally been accepted as the basis of all evolution, but found critical consideration in our specific case recently (Jermy 1984).

Even if one now accepts the view that most secondary plant substances—including the PAs—were in principle also "invented" by the plant against herbivores, one would also expect that some plant feeders "learn" to cope with such plants, for instance by metabolizing the PAs or storing them in an inert

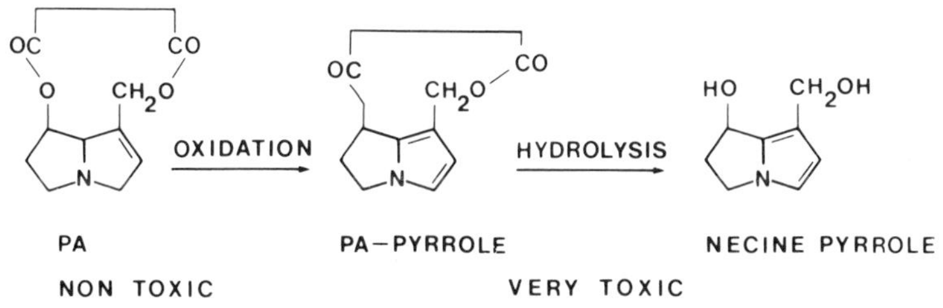

Figure 8.2. Transformation of a nontoxic pyrrolizidine alkaloid (PA) into a toxic pyrrole-alkaloid and finally the necine-pyrrole. This metabolic process is known to occur in the vertebrate liver cells. (See Roitman 1983.)

form. If an animal is able to do this, it immediately enjoys an ideal feeding niche that it shares with only a few equally adapted competitors. Imagine this herbivore to be an insect such as a caterpillar or an aphid; it has now not only plenty of food but also has no risk of being unintentionally eaten by an antelope, as might occur on other plants (Rothschild 1972a, b; Boppré 1978). I will consider this ecological adaptation on unpalatable (toxic) plants as step one of the insect PA–plant relationships, and develop three consecutive steps of such adaptation after this.

Examples of Pyrrolizidine Alkaloid Relations of Insects

Step 1: PA Plants as an Ecological Niche

Insect herbivores that are able to cope with the PA of a relatively protected plant enjoy it as food and shelter.

The larvae of two polyphagous noctuid moth species (*Spodoptera littoralis* and *Melanchra (Mamestra) persicariae*) readily feed on *Senecio vulgaris*, which contains a number of PAs, but chemical analysis revealed that the bodies of both larvae and adults contained no PAs, which presumably had been metabolized (Aplin and Rothschild 1972). Interestingly, these insects are cryptically colored, quite in contrast to those that store the defensive plant substances (see Step 2).

Doubtless, scores of such Step 1 relationships exist between insects and PA plants, but have not been analyzed in detail. In the case of oligophagous insects one can test whether PAs added to their normal food prevents their eating it. For *Locusta migratoria*, a notorious grass feeder, Bernays and Chapman (1977) could "embitter" the normal diet by adding PAs. M. Boppré (unpublished) could quantify the reduction of food uptake of cockroaches, locusts, ants, and of mice by adding PA to the food, and his birds, toads, and lizards rejected food after PA contamination.

Recently a variety of PAs were tested for their deterrence on the larval feeding of the tortricid moth *Choristoneura fumiferana* and only some of them showed a defensive effect. But it remains unknown whether or how those PAs that were eaten were detoxified (Bentley et al. 1984).

Step 2: Transfer of the Protecting Principle (PA) from Plant to Insect and to its Parasite

This is a refinement of step 1, namely that such insects not only feed on the PA plant but store the PA, and thus themselves become unpalatable to their predators.

For my second chapter, another *Senecio* and its specialized consumer insect is an excellent introductory example. *S. jacobaea*, the tansy ragwort, is a temperate-climate weed, spreading over large parts of the world and occupying substantial areas of pasture. Although the plant is avoided by horses and cattle (Figure 8.3) but somewhat consumed by sheep, it is tolerated in small quantities with other plants, but when other food is scarce it is eaten in quantity and this then leads to senecionosis and possibly to loss of the animals. The originally European cinnabar moth, *Tyria jacobaeae* (Arctiidae), feeds as a larva almost exclusively on this plant and all the six PAs that the plant contains (in variable ratios, depending on the season) are also found in the larvae and moths (Aplin and Rothschild 1972). All stages of the moth are brightly—aposematically, warningly—colored and are not eaten by most insectivores, including some arthropods (Windecker 1939). Interestingly, recent experiments with newly hatched domestic chicks showed that the black and yellow ring pattern of *Tyria* caterpillars when painted on meal worms is significantly avoided in comparison with green worms by the naive, young birds. Real *Tyria* larvae were, when pecked at, always released, never eaten (Schuler and Hesse 1985).

Logically, the moth was introduced as a biological weed control measure to some non-European countries where the plant interfered badly with livestock keeping. But the success was not impressive, mainly because of the enormous power of regeneration of this ragwort, which even became a perennial after heavy caterpillar feeding (for references see Myers 1978; Dempster 1982).

Another arctiid, *Nyctemera annulata*, in New Zealand also feeds on ragwort as a larva, stores the PA through its life cycle, and the PA is even found in its parasite. The arctiid is obviously not protected by the PA against the parasite and one wonders whether the parasite now finds the PA a useful defense agent in its own life (Benn et al. 1979). The adaptation to PA plants by the members of this moth family was emphasized some time ago by Rothschild and Aplin (1971).

Figure 8.3. PA-containing plants—here *Senecio Jacobaea*, the tansy ragwort—are avoided by many herbivores as long as other plants are abundant. (Courtesy of M. Boppré.)

Particularly impressive is the interrelationship between PA plants and apo-sematically colored ithomiine butterflies (Nymphalidae) in the New World tropics. Here, the plants are visited by the males, which either ingest the PA-containing nectar of the flowers (which the butterflies pollinate during their visits), or extract it from the surface of the wilting or dry plant surface by ejecting fluid through the proboscis and subsequently sucking the PA solution up again. The PA transfer from the plant to the insect is quite effective since males were found to contain up to 10% of their dry weight as PA after this behavior. The PA was stored and concentrated in the integument, among other places, where the predators can readily sense it (see Boppré 1984a for the earlier literature, and for a recent convincing quantitative report Brown 1984). This latter author also showed that the female ithomiines receive large doses of PA from the males with the spermatophores at mating. After copulation, the female reproductive organs may contain PA at up to 50% of their dry weight. All this probably relates to a highly important sociobiological function of the PA, but with respect to our present consideration it is only worth noting that these PA-containing insects are very well protected. Huge web spiders of the genus *Nephila* are eager to eat butterflies but take ithomiines only before these have consumed PAs; they cut them out of their webs if caught there after the uptake of PAs (Eisner 1982; Brown 1984).

The ithomiine habit of PA uptake from the plant surfaces (often seen by naturalists but not understood until recently) is quite different from that of the feeding cinnabar moth larva, where the automatic uptake of the PA with the *Senecio* leaves is a useful but not nutritious addition. Cases like the Ithomiinae (and some others that will be described below) stimulated Boppré (1984b) to "re-define" this behavior as "pharmacophagy:" "Insects are pharmacophagous if they search for certain secondary plant substances directly, take them up, and utilize them for specific purpose other than primarily metabolism or (merely) foodplant recognition." Clearly, "appetite" of the insects for the PA baits is very impressive for any observer.

Among the moths, species of the arctiid genus *Rhodogastria* behave at night like the ithomiines when they visit wilting PA plants (*Heliotropium* spp.) or PA extracted from such plants, wetting the material with fluid from their pro-boscis and then reimbibing it with the PA (Figure 8.4, Boppré 1981). The moths are joined at these baiting places at night by males of a chrysomelid beetle *(Gabonia)* that often take the "solvent with the PA up from near the tip of the proboscis of the moths" (Boppré and Scherer 1981). *Rhodogastria* is probably protected against predators after PA uptake. When molested, the moth produces bitter, malodorous, and PA-containing (Boppré and Wiedenfeld, unpublished) froth from prothoracic glands. Other arctiids produce such froth but these spe-cies take the juice in again with the proboscis when the danger has passed (Figure 268 in Skaife et al. 1979; Boppré unpublished).

The list of Lepidoptera that come to PA baits by no means ends here, although usually no more than the approach has been observed. The purpose of the PA uptake is still often obscure, but PA protection is probable. Besides Arctiidae, observers have recorded at their baits a number of Ctenuchidae (= Syntomidae),

Figure 8.4. The African arctiid moth *Rhodogastria* sp. is attracted at night by the PA odor of wilting *Heliotropium* sp., dissolves the PA of the plant surface with fluid from its proboscis, and then reimbibes the solution. (See Boppré 1983.)

and, of course, the Danainae, which will be treated in the next section (for references see Pliske 1975; Goss 1979; Boppré 1979, 1981, 1984a).

The final two insects mentioned in this section are two grasshoppers from Africa, both Pyrgomorphidae, known as "poison specialists." They are brilliantly colored, aposematic, and without doubt avoided by most predators. One of them, *Zonocerus variegatus*, feeds on PA plants and stores the PAs, probably for defense (Bernays et al. 1977). The other, *Z. elegans* from South Africa, is close to being truly pharmacophagous, since it not only feeds on withered *Heliotropium*, but also on glass fiber filter paper if it is impregnated with PA (Boppré et al. 1984).

In all the case histories of transfer of the PA from the plant to the animal it has been assumed that the PA survived this procedure with either no or negligible changes. In no case, however, was this unequivocally proved. This might appear to be unnecessary since the biochemical machinery of the insects is said to have difficulties in biosynthesizing PAs de novo and thus all PAs found in insects must stem from plants (Meinwald et al. 1966). Although this may be so with the PAs, any broad generalization is risky. A surprising case became known recently where the plants and their consumers contained the same cyanogenic glucosides (lotaustralin and linamarin) but, in contrast to all expectations, the insects that fed on the plants of the families Passifloraceae and Fabaceae were perfectly able to synthesize these toxic compounds by themselves. The larvae of *Heliconius* and *Zygaena* (Lepidoptera) ate the plants, broke the toxins down metabolically, and resynthesized them, probably during the pupal stage (Nahrstedt and Davis 1983; Wray et al. 1983). This may not be an exceptional case since the production of such glucosides seems to be widespread not only among plants, but also in many insects, even those that never touch plants containing the compounds (C. Naumann, personal communication).

Step 3: PAs as Essential Precursors for the Production of Pheromones

In addition to the use of PAs as protective chemicals, a number of species of three groups of Lepidoptera (Danainae, Ithomiinae and Arctiidae) biosynthesize pheromones from the PAs, which they either acquire by pharmacophagy or larval feeding.

The preferred foodplants of many danaines are the milkweeds (Asclepiadaceae), many of which contain toxic cardiac glycosides (CGs). In one case, the American monarch butterfly *(Danaus plexippus)*, Brower (1969, 1984) produced most impressive proof of the quantitative defensive power of the CGs for this insect. This clear unpalatability of the monarch was then related to the protection that the mimics of this butterfly enjoy. The same CG protection probably exists for some other members of the genus *Danaus* but the assumption that all the danaines are protected by CGs is unjustified (see Boppré 1978, 1984a).

In connection with earlier research on mimicry, the courtship of the queen butterfly *(Danaus gilippus)* was recorded in all its details by Brower et al. (1965). After the female has settled, the male displays his abdominal hairpencils in front of her, and she seems to be persuaded by the odor of this organ and allows copulation. Then came the surprising detection of a pyrrolizine alkaloid as the major component of the odor in this and another danaine species *(Lycorea ceres = L. cleobaea)* by Meinwald and co-workers (1966, 1969). In the following years, two more related volatile compounds were detected in a number of danaines (Figure 8.1:III–V; see Boppré 1984a; Ackery and Vane-Wright 1984).

It is interesting to note that the identification of danaidone in the hairpencils of male danaid butterflies caused the senior chemist of the research team, J. Meinwald (Meinwald and Meinwald 1966), and later Edgar et al. (1971) to indicate the similarity of this substance to the PAs of plants and led to the prediction that the insects might be unable to synthesize such heterocyclic compounds de novo, but rather used plant precursors for this biosynthesis. This was, of course, years before the PA was identified as the pheromone precursor. This "wild" idea was, for several reasons, not readily accepted since the larval danaine foodplants do not, as a rule, contain PAs (Boppré 1978).

Physiological and chemical studies on the old world butterfly *(Danaus chrysippus)* from Australia and Africa then revealed that the adult males need to ingest PAs from dry or wilting PA plants (or even chemically pure PAs in the laboratory) to be able to synthesize the danaidone (Figures 8.5 and 8.6) (Edgar et al. 1973; Schneider et al. 1975). Now, also the mystery of the lack of danaidone in queen butterflies, *D. gilippus*, reared indoors (Pliske and Eisner 1969) was clarified. Incidentally, a most elegant proof for the biological effect of danaidone—and thus for its pheromonal nature!—was brought forward in these experiments. The courtship of the wild males of *D. gilippus* with their danaidone-containing hairpencils is mostly successful, but that of laboratory-reared males, which lack the danaidone, is futile. However, such males from laboratory cultures can be boosted to become successful lovers if one puts danaidone on the hairpencils. Such (close range) mating pheromones have been named arrestants,

Figure 8.5. *Danaus chrysippus* males in the process of PA uptake (as in Figure 8.4) from dry *Heliotropium steudneri* in the East African savannah (left) and in the greenhouse (right).Wingspan, 5.5–7.5 cm. (See Schneider et al. 1975.)

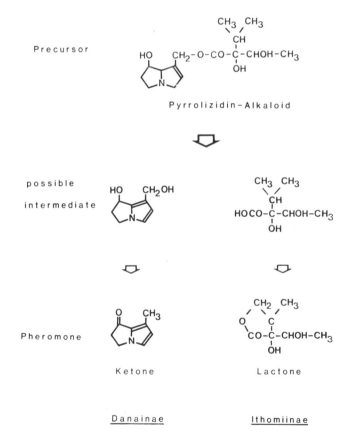

Figure 8.6. Possible intermediates in pheromone biosynthesis from the PA to the ketonic danaidone and the lactonic assembly odor of the ithomiines. (See text.)

aphrodisiacs, etc. (see critical review and all references by Boppré 1984a). As a rule, male pheromones, such as danaidone, are perceived in the Lepidoptera by olfactory receptors on the antennae of both sexes (Schneider and Seibt 1969; Schneider 1984).

In a parallel study it was reported that some ithomiine males produce a pheromone (a lactone) in their wing hairbrushes from the acid-moiety of the PA (Pliske 1975; Pliske et al. 1976; Edgar et al. 1976; Haber 1978). Although the role of the heavy PA load for the protection of these butterflies is now impressively documented by Brown (1984), the role of the pheromone is not yet clear in this group (see Boppré, 1984a).

The PA uptake by Danainae also involves olfactory functions that are the mediators of an upwind search flight leading the insects to the PA source, such as dry parts of *Heliotropium* or a *Crotalaria* seed pod (Schneider et al. 1975). The insects settle on the plant and behave as described before for the Ithomiinae and for *Rhodogastria*. Many field observations of PA plant visits by danaines can now be interpreted in terms of the need for uptake of the pheromone precursor. The appetite for PA by the butterflies is impressive since the African blue monarch male *(Tirumala petiverana)* even scratches and tears the fresh *Heliotropium* leaves, beginning at places where beetles have opened the plant tissue, to gain access to the juice (Figure 8.7) (Boppré 1983). After all this activity the PAs are now in the danaine body but are obviously not metabolized (except for the pheromone production) and will clearly add to the unpalatability of the insect. In a case like the monarch *(D. plexippus)*, the PAs are either additional to the protective power of the CGs or, if the larval foodplant was poor in CG, provide an additional or even exclusive protective chemical. In all those Lepidoptera, including many danaines, that do not eat any larval food with protective chemicals, the PA pharmacophagy is the way to gain protection directly (Boppré 1978, 1984a).

During our experiments, we noticed an additional prerequisite for danaidone production in *D. chrysippus*. We found that biosynthesis occurred only if we allowed the males some "freedom of movement" during sunny mornings. The males then sit with closed wings and thrust their hairpencils, which are still bundled, into their hindwing pouches. These are known from their histology to be glandular in structure, although their function was not understood. After this is done, and only then, is the danaidone biosynthesis finally secured (Boppré et al. 1978). We now know that this or related behavior is effective for such pheromone biosynthesis in the African *Danaus chrysippus*, *Tirumala petiverana* (Figure 8.8), and probably several *Amauris* species (see summarizing reports in Boppré, 1984a and in Ackery and Vane-Wright, 1984). The *Amauris* case is particularly impressive since these species do not have a hindwing pouch or pocket, but an open glandular patch. In their morning business, the males rub a special group of hairs from their hairpencil over the patch where part of the cuticular hairs of the glandular plates breaks off and all this leads to the final danaidone biosynthesis. Similar complex compartmentalization and final mixture of components for a pheromone biosynthesis has been reported from male noctuid moths by Clearwater (1975). However, in the male noctuid moth all this

Figure 8.7. Male danaine, *Tirumala petiverana*, scratching the surface of green *Heliotropium* leaves where alticine beetles had been feeding previously and sucking the fresh plant juice. Wingspan 7–8 cm. (See Boppré 1983.)

Figure 8.8. *Tirumala petiverana* (the African blue monarch butterfly). (Top) Male hairpencils fully expanded, as during courtship or when handled, even by a predator. (Bottom) During "contacting" behavior. The male thrusts each of his closed hairpencils from the inner side into his hindwing pouch, here made visible by cutting the pouch off before the photograph was taken. This contact is the final prerequisite, after the active PA uptake, for danaidone biosynthesis. (See Boppré et al. 1978.)

happens inside the hairbrush apparatus and needs no active outer "behavioral" contact as in the danaines.

With respect to the fate of the PA molecule, many biochemical and histo-chemical questions on these interactions are still open, but the following facts are recorded. First, the majority of the danaines that have been studied contain pyrrolizines in their hairpencils (one exception is the North American monarch *D. plexippus*). Second, we are certain about the PA uptake even with species that could never be experimentally studied because the PA spectrum of the body contains "fingerprints" of PAs as only found in some of the plants that the butterflies must have visited before capture (Edgar et al. 1979). From here on, the biochemical happenings are obscure except that the behavioral contact described above is needed in some species. In others this is still a matter of debate, as in many Euploeini from Asia, although they have wing androconia. In any case biosynthesis must be a "total hairpencil process" in those danaine species that lack any alar androconia such as *Idea* and *Lycorea* (Ackery and Vane-Wright 1984).

Although the ithomiine and the danaine stories are by no means fully under-stood, it is clear that we are observing particularly fascinating cases of plant–insect relationships that must have developed over long periods. On the side of the plants, their protective chemistry is only partially clarified. Some danaine larvae eat CG plants such as *Asclepias,* but Ithomiinae larvae eat solanaceous plants with their specific alkaloids. Both butterfly groups then imbibe PAs for protection and pheromone production as described (quotations see Boppré 1984a). With respect to the question of the evolution, there has been some discussion and speculation on possible deep-rooted early connections between the plant groups providing the food of the two closely related subfamilies of nymphalid butterflies (see Edgar et al. 1974; Edgar 1984; Boppré 1978, 1984a).

Leaving the butterflies, we need to cite some cases of arctiid moths where the larvae feed on PA plants and the male moths have a PA derivative, hy-droxydanaidal (HODal, Figure 8.1) in their abdominal brush organs, which differ widely in structure and size. The first thoroughly studied case was a *Utetheisa* species from North America where the larvae feed relatively exclusively on *Crotalaria* and the pheromone has behavioral effects similar to those seen in the *Danaus* and *Amauris* species (Conner et al. 1981). In other arctiids from East Asia (*Creatonotos gangis* and *C. transiens*), it is again the larva that needs to feed on PA plants or must be supplied with pure PA to ensure that the huge androconial brush organ (the so-called corema) produces HODal (Schneider et al. 1982; Schneider 1983; Bell et al. 1984; Boppré and Schneider 1985a; Wunderer et al. 1986). The function of this organ and its volatile product is rather different from that of *Utetheisa*. We see no quick display in courtship but long-lasting odor dissemination that lures males and later females together into groups (leks?) where matings are observed. HODal obviously is not a rare male odorous com-ponent in related arctiid moth species. We found it recently in *Estigmene ir-regularis* Moore from Sri Lanka (Schulz and Schneider, unpublished), *Pareu-chaetes pseudoinsulata* (Schulz, Kanagaratnam, Francke, Kittmann and Schneider, unpublished), and *"Amsacta" emittens* Walker from Sri Lanka and

South India (Boppré unpublished). Although the larval foodplant of *Estigmene* is not known, *Amsacta* is, like *Creatonotos*, polyphagous, but also inclined to eat PA plants, and *Pareuchaetes* is oligophagous on *Eupatorium* and related PA plants. Because of its foodplant preference this Middle American moth species was introduced to other countries to help control PA-containing weeds, such as *Chromolaena (Eupatorium) odorata* in Sri Lanka (Cock 1984; Cock and Holloway 1982), a strategy similar to the one attempted with the cinnabar moth and *Senecio*.

Step 4: PA-Dependent Growth of a Pheromone Glandular Organ

PA is not only the precursor of a male moth pheromone but also promotes quantitatively the growth of the pheromone producing organ in the pupa: the case of Creatonotos.

Before our study, only *C. gangis* had been observed and reported to have big tubular organs in the male, which the insects display at night for long periods (Pagden 1957; Robinson in Varley 1962). This species and its sister species *C. transiens* were shown to me in 1974 in Sumatra by the Lepidoptera expert Dr. E. W. Diehl at his light trap (Figure 8.9). The first surprise after my return was the detection of HODal in the extracts from the Sumatra samples by the Cornell University chemists (Meinwald and co-workers). This was the second finding

Figure 8.9. *Creatonotos transiens:* beginning and full display (side view, cf. Figure 8.10) of male pheromone dissipating organ, the corema. The odor from this organ first attacts other males that join the beginner and then females that mate with any male that they meet. Wing length, 2 cm. (Courtesy of H. Wunderer.)

of this substance in a moth following the report of Culvenor and Edgar (1972) on *Utetheisa* (Arctiidae), a genus in which several species (like *Creatonotos*) use the same scent, HODal, in the male sex (see Conner et al. 1981 for details).

Since we suspected from our experience with danaines that PAs are also needed as precursor substances for HODal in *Creatonotos*, we fed the larvae of our first laboratory culture in 1978 with either PA-rich plants, or, as control, with PA-free plants (*Heliotropium* and *Senecio* versus *Taraxacum*). In the moths we not only found that the HODal was missing in the *Taraxacum*-fed sample, but also—and this was the surprise—that the scenting organ, the corema, was tiny in the controls (Schneider and Boppré 1981; Schneider et al. 1982; Boppré and Schneider 1985a). Since that time we have done quantitative feeding experiments with both *Creatonotos* species and measured the PA-dependent dimensions of the coremata and the scent hairs (Figure 8.10). The result is a typical exponential growth curve that seems to reach saturation when the larvae were fed 2 mg or more PA. Only the amount of PA and not the larval age is critical, which already allows us to say that the PA is stored (and thus leads to adult unpalatability, Boppré unpublished; see also Pagden 1957). From the size of the corema of a field-caught male we can now "read" in the curve (Figure 8.11) how much of the growth-effective PA its larva ate. This is of interest, since also the field-caught males showed a great variety in the sizes of their coremata (Boppré and Schneider 1985b).

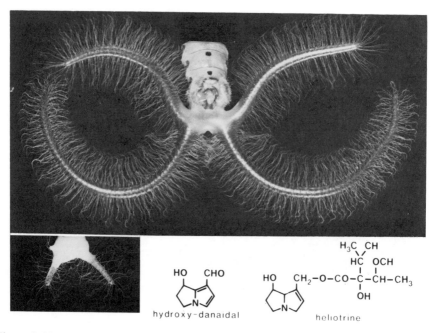

Figure 8.10. *Creatonotos transiens*. PA in the larval diet is the precursor for the pheromone biosynthesis but also controls the pupal growth of the scent-producing organ, the corema. Big coremata seen after ingestion of ample PA in the diet span 35 mm. Small coremata (lower left, same scale) are seen after PA-free feeding. (See Boppré and Schneider 1985b.)

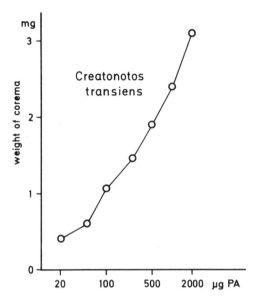

Figure 8.11. The PA in the larval diet of male *Creatonotos (gangis* and *transiens)* regulates the corema growth quantitatively. Ordinate: coremata weight in mg; abscissa: amount of PA fed to the larvae in μg. (cf. Boppré and Schneider 1985b.)

This morphogenetic effect is, as far as we know, restricted to the coremata, organs that grow during pupal life from the genital imaginal disc. In the prepupa only a thickening of the corresponding tissue is seen and this develops in a few days into the huge system of the four tubes. The cells that form the scent hairs in the big organs are, as judged from their nuclear size, highly polyploid. The small coremata have shorter hairs and smaller hair cell nuclei (Egelhaaf and Rick-Wagner, unpublished). The number of hairs ranges from 50 in the smallest case to 3000 hairs in the big organs. The trichogen cells not only "secrete" the hairs, but doubtless are also the site where the HODal biosynthesis takes place.

The critical morphogenetic question, how the PA controls this unique growth in only one organ either directly or indirectly, cannot be answered at the present time. So far we can say that a variety of PAs are equally effective, but some PAs are not. One of our problems in the continuation of these experiments is that few PAs are commercially available.

The HODal biosynthesis is dependent on the PA supply and reaches from zero in the controls to 0.5 mg in the cases where 1–2 mg of PA were fed to the larvae. This corresponds to the highest danaidone amounts in *Danaus chrysippus* (Boppré et al. 1978), of course, dependent on an ample PA supply to the male insect.

In Step 3 we saw that the PA derivatives (Figure 8.1) are definitely pheromones, because they influence the behavior of conspecific individuals during courtship. In the earlier phase of our study we were not certain whether the HODal in *Creatonotos* is such an intraspecies odor signal. Later we learned from cage and greenhouse observations about the peculiar effect of the odor from the coremata in these species (mainly in *C. transiens*). During the early scotophase, males expand their coremata pneumatically (not with hemolymph as in the danaines—they would not even have enough hemolymph to fill the corema tubes and no muscle could retract them quickly, when necessary). This scenting behavior attracts first other males that might join the first perfumer, and then females come and readily mate with the first male they meet. Later, the male calling subsides and now the females begin to rhythmically puff their effective odor bouquet out, which attracts males and leads to more matings as in most moths (Schneider 1983; Wunderer et al. in press; Boppré and Schneider 1985a).

The striking formation of a group of scenting males to which females are also attracted is reminiscent of a lek, or an assembly like a tournament, where the ladies have the choice (for the fittest!). Unfortunately the *Creatonotos* females do not seem to select a specific male, for instance, the one with the biggest corema and strongest perfume, but we must confess that our indoor experiments might be misleading. Whatever the final interpretation of this intriguing behavior will be, we may be sure that this PA-derived odor is also a pheromone, which we even found to be "served" by the enantiomer specific receptor cells on the antennae of both sexes of *Creatonotos* (Wunderer et al. in press; Bell et al. 1984).

Synopsis

What now is the "strange fate" of the PAs? They are, as all complex products of the anabolic organismic machinery, eventually catabolized to simpler molecules, or to atoms. But during their existence, the PA molecules first of all serve as defensive elements for their producers, the plants. Specialized plant eaters now overcome the defense and some even find in the PAs welcome substances that they keep and store and that now also protect them from being eaten.

If a plant dies, it wilts and dries and then some of its contents become accessible to adult Lepidoptera, which have only soft mouthparts, the proboscis, for fluid uptake. They find the wilting plant with their "nose," the antenna, and suck in the dissolved PA. As compared to normal feeding this is an unusual case because when the butterflies and moths (as well as beetles and grasshoppers) do this, they are not interested in any uptake of nutrients, but want the PA as a kind of pharmacon; they are pharmacophagous.

In both groups, the PA herbivores and the PA pharmacophages, we find cases where the PA molecule became a building stone for a chemical signal, a pheromone, and at least this step might be called strange. The arctiid moths and the milkweed butterflies (danaines) both "invented" their pyrrolizidine pheromones from the PAs and one wonders what it was that they both (probably independently) found so useful for this biogenesis. Did they perhaps do this since some of their ancestors detected the defensive power of the PA first? In the third case of pheromone production from PA, in the ithomiine butterflies, it is the acidic moiety of the PA that is used for the construction of the pheromone. Could this by any chance be a particularly elegant way to keep the protective pyrrolic moiety and make the pheromone from the other? In all these cases, the available quantity of the PA limits the pheromone production.

References

Ackery PR, Vane-Wright RI (1984) Milkweed Butterflies. British Museum of Natural History, London

Aplin RT, Rothschild M (1972) Poisonous alkaloids in the body tissues of the Garden Tiger moth [*Arctia caja* (L.)] and the Cinnabar moth [*Tyria* (= *Callimorpha*) *jacobaeae* (L.)] (Lepidoptera). In: de Vries A, Kochva E (eds) Toxins of Animal and Plant Origin 2. Gordon and Breach, London, pp 579–595

Bell TW, Boppré M, Schneider D, Meinwald J (1984) Stereochemical course of pheromone biosynthesis in an arctiid moth *(Creatonotos transiens)*. Experientia 40:713–714

Benn M, DeGrave J, Gnanasundersam C, Hutchins R (1979) Host-plant pyrrolizidine alkaloids in *Nyctemera annulata* Boisduval: their persistence through the life-cycle and transfer to a parasite. Experientia 35:731–732

Bentley MD, Leonard DE, Stoddard WF, Zalkow LH (1984) Pyrrolizidine alkaloids as larval feeding deterrents for spruce budworm *Choristoneura fumiferana* (Lepidoptera: Tortricidae). Ann Entomol Soc Am 77:393–397

Bernays EA (1983) Nitrogen in defence against insects. In: Lee JA, McNeill S, Rorison IH (eds) Nitrogen as an Ecological Factor. Blackwell, Oxford, pp 321–344

Bernays EA, Chapman RF (1977) Deterrent chemicals as a basis of oligophagy in *Locusta migratoria* (L.). Ecol Entomol 2:1–18

Bernays EA, Edgar JA, Rothschild M (1977) Pyrrolizidine alkaloids sequestered and stored by the aposematic grasshopper, *Zonocerus variegatus*. J Zool Lond 182:85–87

Boppré M (1978) Chemical communication, plant relationships, and mimicry in the evolution of danaid butterflies. Entomol Exp Appl 24:264–277

Boppré M (1979) Lepidoptera and withered plants. Antenna 3:7–9

Boppré M (1981) Adult Lepidoptera "feeding" at withered *Heliotropium* plants (Boraginaceae) in East Africa. Ecol Entomol 6:449–452

Boppré M (1983) Leaf-scratching—a specialized behaviour of danaine butterflies for gathering secondary plant substances. Oecologia 59:414–416

Boppré M (1984a) Chemically mediated interactions between butterflies. In: Vane-Wright RI, Ackery PR (eds) The Biology of Butterflies. Academic Press, London, pp 259–275

Boppré M (1984b) Redefining "pharmacophagy." J Chem Ecol 10:1151–1154

Boppré M, Scherer G (1981) A new species of flea beetle (Alticinae) showing male-biased feeding at withered *Heliotropium* plants. Syst Entomol 6:347–354

Boppré M, Schneider D (1985a) On the biology of *Creatonotos* (Lep.: Arctiidae) with special reference to the androconial system. Zool J Linn Soc (in press)

Boppré M, Schneider D (1985b) Pyrrolizidine alkaloids quantitatively regulate both scent organ morphogenesis and pheromone biosynthesis in male *Creatonotos* moths (Lepidoptera: Arctiidae). J Comp Physiol 157A:569–577

Boppré M, Petty RL, Schneider D, Meinwald J (1978) Behaviorally mediated contacts between scent organs: another prerequisite for pheromone production in *Danaus chrysippus* males (Lepidoptera). J Comp Physiol 126:97–103

Boppré M, Seibt U, Wickler W (1984) Pharmacophagy in grasshoppers? Entomol Exp Appl 35:115–117

Brower LP (1969) Ecological chemistry. Sci Am 220 (2):22–29

Brower LP (1984) Chemical defense in butterflies. In: Vane-Wright RI, Ackery PR (eds) The Biology of Butterflies. Academic Press, London, pp 109–134

Brower LP, Brower JVZ, Cranston FP (1965) Courtship behavior of the queen butterfly, *Danaus gilippus berenice*. Zoologica NY 50:1–39

Brown KS Jr (1984) Adult-obtained pyrrolizidine alkaloids defend ithomiine butterflies against a spider predator. Nature 309:707–709

Bull LB, Culvenor CCJ, Dick AT (1968) The Pyrrolizidine Alkaloids. North-Holland Publishing Co., Amsterdam

Clearwater JR (1975) Pheromone metabolism in male *Pseudaletia separata* (Walk) and *Mamestra configurata* (Walk), (Lepidoptera, Noctuidae). Comp Biochem Physiol B 50:77–82

Cock MJW (1984) Possibilities for biological control of *Chromolaena odorata*. Trop Pest Manage 30:7–13

Cock MJW, Holloway JD (1982) The history of, and prospects for, the biological control of *Chromolaena odorata* (Compositae) by *Pareuchaetes pseudoinsulata* Rego Barros and allies (Lepidoptera: Arctiidae). Bull Entomol Res 72:193–205

Conner WE, Eisner T, Vander Meer RK, Guerrero A, Meinwald J (1981) Precopulatory sexual interaction in an arctiid moth *(Utetheisa ornatrix):* role of a pheromone derived from dietary alkaloids. Behav Ecol Sociobiol 9:227–235

Culvenor CCJ, Edgar JA (1972) Dihydropyrrolizine secretions associated with coremata of *Utetheisa* moths (family Arctiidae). Experientia 28:627–628

Dempster JP (1982) The ecology of the cinnabar moth, *Tyria jacobaeae* L. (Lepidoptera: Arctiidae). Adv Ecol Res 12:1–36

Edgar JA (1984) Parsonsieae: ancestral larvae foodplants of the Danainae and Ithomiinae. In: Vane-Wright RI, Ackery PR (eds) The Biology of Butterflies. Academic Press, London, pp 91–93

Edgar JA, Culvenor CCJ, Smith LW (1971) Dihydropyrrolizine derivatives in the "hairpencil" secretions of danaid butterflies. Experientia 27:761–762

Edgar JA, Culvenor CCJ, Robinson GS (1973) Hairpencil dihydropyrrolizines of Danainae from the New Hebrides. J Aust Entomol Soc 12:144–150

Edgar JA, Culvenor CCJ, Pliske TE (1974) Co-evolution of danaid butterflies and their hostplants. Nature 250:646–648

Edgar JA, Culvenor CCJ, Pliske TE (1976) Pyrrolizidine alkaloid-derived pheromone on the costal fringes of male Ithomiinae. J Chem Ecol 2:263–270

Edgar JA, Boppré M, Schneider D (1979) Pyrrolizidine alkaloid storage in African and Australian danaid butterflies. Experientia 35:1447–1448

Ehrlich PR, Raven PH (1967) Butterflies and plants: a study in coevolution. Evolution 18:586–608

Eisner T (1982) For love of nature: explorations and discovery at biological field stations. Bioscience 32:321–326

Fraenkel G (1959) The raison d'être of secondary plant substances. Science 129:1466–1470

Goss GJ (1979) The interaction between moths and plants containing pyrrolizidine alkaloids. Environ Entomol 8:487–493

Haber WA (1978) Evolutionary ecology of tropical mimetic butterflies. Ph.D. Thesis University of Minnesota

Huxtable RJ (1980) Herbal teas and toxins: novel aspects of pyrrolizidine poisoning in the United States. Perspect Biol Med 24:1–14

Jermy T (1984) Evolution of insect/host plant relationships. Am Natur 124:609–630

Lüthy J, Heim T, Schlatter C (1983) Transfer of (^{3}H)pyrrolizidine alkaloids from *Senecio vulgaris* L. and metabolites into rat milk and tissues. Toxicol Lett 17:283–288

Meinwald J, Meinwald YC (1966) Structure and synthesis of the major component in the hairpencil secretion of a male butterfly, *Lycorea ceres ceres* (Cramer). J Am Chem Soc 88:1305–1310

Meinwald J, Meinwald YC, Wheeler JW, Eisner T, Brower LP (1966) Major components in the exocrine secretion of a male butterfly *(Lycorea)*. Science 151:583–585

Meinwald J, Meinwald YC, Mazzocchi PH (1969) Sex pheromone of the queen butterfly. Science 164:1174–1175

Mothes K (1980) Historical introduction. In: Bell EA, Charlwood BV (eds) Encyclopedia of Plant Physiology. New Ser, 8. Secondary plant products. Springer-Verlag, Berlin, pp 1–10

Mothes K (1984) Zur Wissenschaftsgeschichte der biogenen Arzneistoffe. In: Czygan FC (ed) Biogene Arzneistoffe. Vieweg, Braunschweig-Wiesbaden, pp 5–25

Myers JH (1978) Biological control introductions as grandiose field experiments: adaptations of the cinnabar moth to new surroundings. 4th Int Symp Biol Contr Weeds, 181–188

Nahrstedt A, Davis RH (1983) Occurrence, variation and biosynthesis of the cyanogenic glucosides linamarin and lotaustralin in species of the *Heliconiini* (Insecta: Lepidoptera). Comp Biochem Physiol B 75:65–73

Pagden HT (1957) The presence of coremata in *Creatonotos gangis* (L.) (Lepidoptera: Arctiidae). Proc R Entomol Soc Lond A 32:90–94

Pliske TE (1975) Attraction of Lepidoptera to plants containing pyrrolizidine alkaloids. Environ Entomol 4:455–473

Pliske TE, Eisner T (1969) Sex pheromone of the queen butterfly: biology. Science 164:1170–1172

Pliske TE, Edgar JA, Culvenor CCJ (1976) The chemical basis of attraction of ithomiine butterflies to plants containing pyrrolizidine alkaloids. J Chem Ecol 2:155–162

Röder E (1984) Wie verbreitet und wie gefährlich sind Pyrrolizidin Alkaloide? Pharm Unserer Zeit 13 (2):33–38

Roitman JN (1983) Ingestion of pyrrolizidine alkaloids: a health hazard of global proportions. In: Finley JW, Schwass DE (eds) Xenobiotics in Foods and Feeds. Am Chem Soc, Washington DC, pp 345–378

Rothschild M (1972a) Secondary plant substances and warning coloration in insects. In: van Emden HF (ed) Insect/Plant Relationships. Blackwell, Oxford, pp 59–83

Rothschild M (1972b) Some observations on the relationship between plants, toxic insects and birds. In Harborne JB (ed) Phytochemical Ecology. Academic Press, London, pp 1–12

Rothschild M, Aplin RT (1971) Toxins in tiger moths (Arctiidae: Lepidoptera). In Tahori AS (ed) Pesticide Chemistry 3. Chemical Releasers in Insects. Gordon and Breach, London, pp 177–182

Schneider D (1983) Kommunikation durch chemische Signale bei Inskten: alte und neue Beispiele von Lepidopteren. Verh Dtsch Zool Ges 1983:5–16

Schneider D (1984) Pheromone biology in the Lepidoptera: overview, some recent findings and some generalizations. In: Bolis L, Keynes RD, Maddrell SHP (eds) Comparative Physiology of Sensory Systems. Cambridge University Press, pp 301–313

Schneider D, Boppré M (1981) Pyrrolizidin-Alkaloide als Vorstufen für die Duftstoff-Biosynthese und als Regulatoren der Duftorgan-Morphogenese bei *Creatonotos* (Lepidoptera: Arctiidae). Verh Dtsch Zool Ges 1981:169

Schneider D, Seibt U (1969) Sex pheromone of the queen butterfly: electroantennogram responses. Science 164:1173–1174

Schneider D, Boppré M, Schneider H, Thompson WR, Boriack CJ, Petty RL, Meinwald J (1975) A pheromone precursor and its uptake in male *Danaus* butterflies. J Comp Physiol 97:245–256

Schneider D, Boppré M, Zweig J, Horsley SB, Bell TW, Meinwald J, Hansen K, Diehl EW (1982) Scent organ development in *Creatonotos* moths: regulation by pyrrolizidine alkaloids. Science 215:1264–1265

Schuler W, Hesse E (1985) On the function of warning coloration: a black and yellow pattern inhibits prey-attack by naive domestic chicks. Behav Ecol Sociobiol 16:249–255

Skaife SH, Ledger JI, Bannister A (1979) African Insect Life. C Stuik Publ, Cape Town

Swain T (1976) Secondary compounds: primary products. Nova Acta Leopold Suppl 7:411–421

Teuscher E (1984) Zur möglichen Funktion von Sekundärstoffen in biologischen Systemen In: Czygan FC (ed) Biogene Arzneistoffe. Vieweg, Braunschweig-Wiesbaden, pp 61–83

Varley CG (1962) A plea for a new look at Lepidoptera with special reference to the scent distributing organs of male moths. Trans Soc Br Entomol 15:29–40

Windecker W (1939) *Euchelia jacobaea* und das Schutztrachtenproblem. Z Morphol Oekol Tiere 35:84–139

Wray V, Davis RH, Nahrstedt A (1983) Biosynthesis of cyanogenic glycosides in butterflies and moths: incorporation of valine and isoleucine into linamarin and lotaustralin by *Zygaena* and *Heliconius* species (Lepidoptera). Z Naturforsch 38 C:583–588

Wunderer H, Hansen K, Bell TW, Schneider D, Meinwald J (1986) Male and female pheromones in the behaviour of two Asian moths, *Creatonotos* (Lepidoptera: Arctiidae). Exp Biol 46 (in press)

Chapter 9

The Role of Experience in the Host Selection of Phytophagous Insects

TIBOR JERMY*

There is strong observational and experimental evidence that in phytophagous insects host plant recognition and utilization on the one hand, and avoidance or rejection of nonhost plants, on the other hand, are largely determined genetically, thus, they cannot be changed basically by experience. Nevertheless, it has been shown, mostly during the last two decades, that experience may play a considerable role in shaping subtle details of host plant finding, acceptance, preference, etc., from which important general conclusions can be drawn.

In the following an attempt is made to survey briefly our present knowledge of the main phenomena revealed so far in this connection, namely (1) induced host preferences, concerning both feeding and oviposition; (2) habituation to feeding inhibitory stimuli; and (3) food aversion learning.

Induced Feeding Preferences

It has been found repeatedly that the plant-feeding instars of various phytophagous insect species, when ingesting a particular food for some time, afterwards show increased preference for that food over others that beforehand were equally or even more acceptable (see Jermy et al. 1968 for references). This phenomenon was given the term "induced preference" by Jermy et al. (1968), who made the first detailed investigations into it. As Dethier (1982) pointed out this neutral term should be preserved because the nature of induced preference is largely unknown so it does not fit into the usual categories of learning.

*Institute for Plant Protection, Hungarian Academy of Sciences, Budapest, Pf. 102. H-1525, Hungary.

Occurrence of Induced Feeding Preferences

Table 9.1 contains a list of insect species in which feeding preference induction has been found so far. Although the data are still relatively scarce, it can be concluded that induction of feeding preferences is not restricted to certain taxonomic groups.

Lack of induction has been reported rarely, presumably also because negative results are less often published. For example, although de Wilde et al. (1960) first found experience-induced changes in host plant selection with *Leptinotarsa decemlineata* adults, later experiments carried out in the same laboratory did not confirm these findings (Bongers 1965). Negative results were obtained with *Pieris rapae* and *P. napi macdunnoughii* (Chew 1980), and with *Hyphantria cunea* and *Mamestra brassicae* (Jermy et al. unpublished). However, the fact that not all host plant species are equally suitable for induction (see below) may partly be responsible for negative results.

Behavioral Steps Involved

The behavioral steps involved in the changes in food preferences include both orientation and feeding, as has been shown by Saxena and Schoonhoven (1978, 1982). As a result, preference induction is manifest to a greater degree in choice tests than in no-choice tests, because orientation is involved in the former, but not in the latter. This indicates the importance of orientation, i.e., olfaction, in the process of food choice.

The Strength of Induction

The degree of preference for the novel food in experienced insects compared with that in inexperienced ones as well as the rigidity of the novel behavior vary considerably from plant to plant and from insect species to species.

In this connection Yamamoto (1974) and Flowers and Yamamoto (1982) assumed that even oligophagy in the solanaceous feeder, *Manduca sexta*, is itself due to induction, because first-instar larvae accepted some nonsolanaceous plants rejected by later instars that had been reared on solanaceous species. However, de Boer and Hanson (1984) have shown that the relative preferences

Table 9.1. List of insect species in which induction of feeding preferences has been shown experimentally

| | Stages | |
Species	tested	References
Phasmatodea		
Carausius morosus	L, A	Cassidy 1978
Heteroptera		
Dysdercus koenigi	L	Saxena 1967

Table 9.1. *Continued*

Species	Stages tested	References
Homoptera		
Acyrthosiphon pisum	A	Hubert-Dahl 1975
Schizaphis graminum	A	Schweissing and Wilde 1979
Coleoptera		
Epilachna pustulosa	A	Iwao 1959
Subcoccinella 24-punctata	L, A	Ali 1976
Haltica lythri	L, A	Phillips 1977
Galerucella lineola	L, A	Kozhančikov 1958
Lepidoptera		
Noctuidae		
Heliothis armigera	L	Aboul-Nasr et al. 1981
Heliothis zea	L	Jermy et al. 1968
	L	Wiseman and McMillian 1980
Spodoptera eridania	L	Scriber 1982
Lymantriidae		
Euproctis chrysorrhoea	L	Kuznetzov 1952
	L	Getzova and Lozina-Lozinskii 1955
Lymantria dispar	L	Barbosa et al. 1979
	L	Getzova and Lozina-Lozinskii 1955
	L	Wasserman 1982
Arctiidae		
Hyphantria cunea	L	Greenblatt et al., 1978
Sphingidae		
Manduca sexta	L	Jermy et al. 1968
	L	Hanson and Dethier 1973
	L	Yamamoto 1974
	L	Flowers and Yamamoto 1982
	L	Saxena and Schoonhoven 1982
	L	de Boer and Hanson 1984
Saturniidae		
Antheraea pernyi	L	Getzova and Lozina-Lozinskii 1955
Antheraea polyphemus	L	Hanson 1976
Callosamia promethea	L	Hanson 1976
Hyalophora cecropia	L	Grabstein and Scriber 1982b
Limenitis archippus	L	Hanson 1976
Limenitis astyanax	L	Hanson 1976
Limenitis hybrid *rubidus*	L	Hanson 1976
Pieridae		
Pieris brassicae	L	Johansson 1951
		Ma 1972
Papilionidae		
Papilio aegeus	L	Stride and Straatman 1962
Papilio glauca	L	de Boer and Hanson 1984
Papilio machaon	L	Wiklund 1973
Nymphalidae		
Chlosyne lacinia	L	Ting 1970
Polygonia interrogationis	L	Hanson 1976

for host over acceptable nonhost plant species were maintained by rearing on the former plant species and were reduced by rearing on the latter. They concluded that oligophagy in *M. sexta* is inherited and not induced.

In general, inexperienced first-instar lepidopterous larvae are less restricted than later stages in their food choice but this should not be regarded as polyphagy since they also show a genetically fixed range of plant species that they may accept more or less (Wiklund 1973; de Boer and Hanson 1984). Controversies arising in this connection are mainly conceptual. Namely, with oligophagous insects, especially with respect to induction, it is necessary to distinguish three groups of plants as did de Boer and Hanson (1984): (1) host plant species that are attacked in nature; (2) acceptable nonhost plants that are not attacked in nature but are acceptable and incidentally also suitable for normal development under experimental conditions; and (3) unacceptable nonhost plants that are fully rejected under all conditions. To which group a given plant species belongs may depend, among other things, on varietal differences and on growing conditions that determine plant quality.

It has been demonstrated that preference can be induced for the first two groups of plant species, i.e., within the host plant range sensu lato, but not for unacceptable nonhost plants (Jermy et al. 1968; de Boer and Hanson 1982). The latter authors have shown that, in general, the probability and strength of induction in *Manduca sexta* was inversely proportional to the taxonomic relatedness of the plant species paired in the choice tests.

Since oligophagous insects can be tested ab ovo only with closely related plant species, one could presume a negative correlation between the degree of host plant specificity and the probability of induction. However, de Boer and Hanson (1984) found no such correlation from a survey of the data on lepidopterous species. It has to be considered, however, that plant taxonomic relationships do not necessarily reflect the relationships of the "chemosensory profiles" that an insect perceives.

Persistence (rigidity) of induced preference has been studied in a few cases. Jermy et al. (1968) found that induced preference for some plants in *Heliothis zea* may persist through two molts and one whole instar when the latter was fed with an artificial diet. Cassidy (1978) has shown that in *Carausius morosus* induced preference can be extinguished by feeding on another food plant, although this flexibility decreased with age, the adults showing very little alteration of initial preference. The rigidity of induction in *Pieris brassicae* is striking: when larvae were reared on *Brassica oleracea* and the young fifth-instar larvae were transferred to another host plant, *Tropaeolum majus*, all larvae died of lack of food intake (Ma 1972)! Hanson (1976) found the same with fifth-instar larvae of *Callosamia promethea* and Scriber (1982) with *Spodoptera eridania*. This may be called the "starving-to-death-at-Lucullian-banquets" phenomenon. We shall come back to this later.

Strong and rigid induction thus may make it very difficult to discriminate between behavioral (preingestive) and physiological (postingestive) effects of food on an insect (Grabstein and Scriber 1982a). It also indicates that great care has to be taken when the possible host plant range of an insect has to be determined, e.g., at the introduction of phytophagous insects for weed control.

Individual differences in insects concerning the strength of induction have been shown repeatedly (Jermy et al. 1968; Ma 1972; Phillips 1977; Cassidy 1978). Whether this is due to genetically determined variability remains to be clarified for each case.

Time Needed for Induction

Very little is known about the minimum time needed for induction. Strong induction has been observed in *Heliothis zea* after only 24 hr of feeding on an inducing plant (Jermy et al. 1968). Ma (1972) demonstrated significant induction in *Tropaeolum*-reared *Pieris brassicae* larvae after 4 hr of feeding although 2 hr were not enough to overcome preferences already established.

The Constituents of Food Involved in Induction

This has been studied by Städler and Hanson (1978), who demonstrated that in larvae of *Manduca sexta* artificial diets may also induce preferences and that nutrients, especially lipid components, provided the major portion of information used in food discrimination and induction, although aqueous fractions also played a role. Saxena and Schoonhoven (1978) were able to induce orientational and feeding preference to an artificial diet that differed from the control diet only in the presence of a single odorous compound (citral). But since the plants contain an array of compounds to which the insects may respond behaviorally, perhaps many plant constituents are involved in preference induction.

Neural Mechanisms Underlying Induction

These have not been studied extensively. Dethier (1982) in his excellent survey of mechanisms of host plant recognition rightly emphasized (p 49) that

> Insofar as host-plant recognition by insects is concerned we are in total ignorance of the nature of gustatory and olfactory neural pathways and events.

Since plant recognition is the basis of preference induction, Dethier's statement applies to the neural mechanisms of induction, too. Notwithstanding this fact, there are important findings in this field that may help to create a more complete picture of the entire phenomenon in the future.

All authors studying the sensory background of feeding preference induction agree that chemical information perceived by taste receptors plays a decisive role, although olfaction is also involved to a lesser degree. In the most intensively studied lepidopterous larvae the maxillary sensilla styloconica and receptors of the preoral cavity were found to be the most important gustatory receptors whereas the gustatory function of the palpi was insignificant. The palpi and the antennae as olfactory organs were found to play a less important role in induction (Ma 1972; Hanson and Dethier 1973). Nevertheless, Saxena and Schoonhoven (1982) demonstrated by behavioral experiments with *Manduca sexta* larvae that orientational (olfactory) preferences induced by the rearing food do play a role in enabling the insect to reach a certain food and thus may, in concert with taste preferences, determine the outcome of choice tests.

The question whether induction of preferences involves central and/or peripheral modifications in the nervous system has been investigated by several authors. Schoonhoven (1969) found that feeding on a diet containing the deterrent compound salicin later reduced the behavioral response to this compound. Since at the same time the electrophysiologically determinable neural response also decreased, it was suggested that induction of preference may be due to reduction in sensitivity to deterrents. On the other hand, Städler and Hanson (1976) have shown that induction increased neural responses to phagostimulating extracts of the inducing food. Thus, both findings indicated the importance of peripheral changes during feeding. However, the fact that at least one molt in *Pieris brassicae* (Ma 1972) and two molts in *Heliothis zea* (Jermy et al. 1968) did not extinguish induced preferences seems to prove the leading role of the central nervous system (Jermy et al. 1968), although the possibility of the occurrence of both peripheral and central changes cannot be excluded (Schoonhoven 1977).

One of the main conclusions that can be drawn from the results of investigations into feeding preference induction is that they clearly prove the insects' ability to discriminate sharply not only between host and nonhost plants but also between equally suitable host plants. Thus, the insects must be able to perceive a very detailed "chemosensory profile" (de Boer and Hanson 1984), i.e., a certain chemical "Gestalt" (Kogan 1977) of each plant species. This strongly supports Dethier's (1973) conclusions on the complexity of stimuli involved in host recognition by phytophagous insects.

Induced Oviposition Preferences

Although relatively many studies have been conducted on the induction of feeding preferences, investigations into the changes in oviposition behavior caused by experience remain quite scarce. Induced oviposition preferences have been observed in Lepidoptera, Coleoptera, and Diptera species.

Lepidoptera

On the basis of insectary and field observations Gilbert (1975) assumed that females of the butterflies *Heliconius* spp. associate the shape of the host plant leaves with their chemistry and afterwards searched for the same leaf shape.

Wiklund (1982) found the same behavior in *Papilio machaon*, but the butterflies *Leptidea sinapis*, *Fabriciana adippe*, and *Lycaena phalaeas* were found to be unable to recognize a host plant in flight. Rausher (1978) demonstrated that ovipositing females of the swallowtail butterfly, *Battus philenor*, search selectively for either broad- or narrow-leaved larval host plants, indicating the use of a search image (or better: searching image, McFarland 1981) that is learned. Some females were found to switch preference from one leaf shape to another after relatively little experience.

Traynier (1979) studied the influence of the chemical stimuli received by tarsal contact with a host plant leaf on the subsequent oviposition behavior of *Pieris rapae* females. These stimuli increased the tendency of the females to approach and land on substrates depending on previous experience. Since the changes were immediate and long-term and were not connected with the physiological state of the ovaries, it was assumed that changes in the CNS were involved.

No induction could be shown with the pierid butterfly *Colias eurytheme* (Tabashnik et al. 1981).

Coleoptera

Mark (1982) demonstrated induced oviposition preferences in the bruchid *Callosobruchus maculatus* as a result of direct adult experience but not due to larval experience.

Diptera

Detailed investigations were conducted into experience-induced changes in the oviposition behavior of the apple maggot fly, *Rhagoletis pomonella*. Prokopy (1977) found that the attraction of egg-laying females to large spheres decreased as the season progressed and suggested that this was due to learning. Experiments with fruits indicated that previous experience on apple or hawthorn fruits induced oviposition preference for the respective fruit plant species, more or less modifying the genetically based propensities (Prokopy et al. 1982a, b). It has been assumed that in this fly both types of learning occur, namely, "learning to reject" the novel fruit and "learning to accept" the familiar fruit. The former type dominates in the females after they arrive on a fruit whereas the latter is dominant during the fruit-finding phase of foraging behavior (Prokopy et al. 1986).

With a few exceptions (e.g., Smith and Cornell 1979) there is general agreement among authors that, contrary to the Hopkins' Host Selection Principle (Hopkins 1917), larval experience does not influence the females' oviposition behavior (for references and results see Jermy et al. 1968; Wiklund 1974; Stanton 1979; Fox and Morrow 1981; Tabashnik et al. 1981; Mark 1982; Prokopy et al. 1982b, 1986).

The number of insect species studied so far does not allow any generalization concerning the incidence of induced oviposition preferences in different taxonomic groups. The neural mechanisms underlying induction are totally unknown. On the basis of studies on the stimuli involved in oviposition site selection one can only speculate that in induction of oviposition preferences receptors located on the head appendages and/or on the tarsi play the leading role whereas those on the ovipositor are less important (Jermy and Szentesi 1978).

Oviposition in most insects is a much more complex process than feeding since in many cases navigation and the association of visual and chemical information is involved. This also implies that investigations into induced oviposition preferences are methodologically much more difficult than studies on food preference induction.

Habituation to Feeding Inhibitory Stimuli

Habituation has been defined as the waning of the response to a repeatedly presented stimulus (Thompson and Spencer 1966). With insects it has been reported that previous experience with a food containing feeding inhibitors may increase the acceptance of that food during subsequent exposure (Strebl 1928; Schoonhoven 1969; Gill 1972). Field experiments also indicated that antifeedants sprayed on plants to protect them against insect pests often lost their effectiveness after a relatively short time (Bernays 1983).

Experiments to reveal the occurrence and the mechanism of behavioral habituation to deterrents have been carried out recently on the larvae of two oligophagous insects, *Locusta migratoria* and *Pieris brassicae,* and two polyphagous ones, *Schistocerca gregaria* and *Mamestra brassicae* (Jermy et al. 1982; Szentesi and Bernays 1984). In these experiments food treated with secondary plant substances having feeding inhibitory effects was given to the insects for a limited period each day. After several days the acceptance of the treated food by experienced larvae, compared with the acceptance shown by the inexperienced, naive, ones, was determined. In several instances an increase of the acceptance of deterrent-treated food by the experienced larvae was found. This has been regarded as behavioral habituation to the deterrents. So far, polyphagous insects seem better able to habituate than oligophagous ones.

Habituation occurred only to relatively weak feeding inhibitory stimuli. Concentrations of deterrent chemicals totally inhibiting feeding for about 12 hr did not elicit habituation (Jermy et al. 1982; Szentesi and Bernays 1984). This agrees with the generally accepted assumption that weaker stimuli are more effective in eliciting habituation than stronger ones (Thompson and Spencer 1966; Hinde 1970). However, in recent preliminary experiments with nonhost plants, no habituation was found even if the plant's inhibitory effect was relatively weak. This may suggest that the insects cannot habituate to the complex of substances that inhibit feeding on nonhost plants (Jermy et al., unpublished).

The occurrence of habituation did not depend on the type of deterrent secondary plant substances. However, the same chemical may evoke different responses depending on the overall stimulus situation. For example, nicotine hydrogen tartrate, when presented on leaf material, produced habituation in the larvae of *Schistocerca gregaria,* but when it was added to sucrose-impregnated glass fiber paper, which is a very poor diet, the opposite phenomenon was observed: the experienced insects ate less than the naive ones. This was regarded as aversion learning, although there was no noticeable evidence of a toxic effect which is supposed to be the prerequisite for this type of learning (Jermy et al. 1982).

To reveal the sensory and physiological feedback leading to habituation, Szentesi and Bernays (1984) fitted small nylon capillaries over each maxillary palp of fifth-instar *Schistocerca gregaria* larvae and filled the capillaries on four consecutive days with the deterrent nicotine hydrogen tartrate solution for 30, 30, 120, and 150 min, respectively. Control insects were treated in the same way with water. In the feeding assay, at the end of the experiment, the empty capillaries were left on the palps to prevent stimulation from the assay diet. The insects whose palps were previously exposed to the deterrent solution showed habituation whereas the controls did not. This indicates that sensory adaptation was not the basis of habituation and that the CNS was involved. This experiment also demonstrated that a postingestional physiological effect was not a sine qua non of habituation. Nevertheless, the neural basis of habituation remains largely unknown. It is also a controversial issue whether habituation is a nonassociative process, and thus not true learning (Marlin and Miller 1981), or whether the insects do "associate" the deterrent with an otherwise suitable food (Szentesi and Bernays 1984).

Great individual variation has been observed in all habituation experiments. How far this is due to individual differences of the insects' physiological state, neural characters, or to differences in response to handling during experimentation is unknown.

It is clear that many more investigations are needed before the physiological and ecological importance of habituation to deterrents can be understood. More extensive knowledge in this field may have relevance to selecting plant cultivars whose insect resistance is based on deterrents and to applying antifeedants for pest control, as well as to understanding changes in the host plant range of natural insect populations.

Food Aversion Learning

Dethier (1980) was the first to demonstrate food aversion learning in insects, namely, in the larvae of the polyphagous arctiid species *Diacrisia virginica* and *Estigmene congrua*. In both species the palatable plant *Petunia hybrida* caused illness and after recovery the larvae rejected this plant in the choice test, indicating the association of illness with the chemosensory image of *Petunia*. As mentioned above, a kind of aversion learning has been found also in the polyphagous *Schistocerca gregaria* (Jermy et al. 1982).

On the other hand, Dethier and Yost (1979) did not find aversion learning in the oligophagous sphingid *Manduca sexta* which is in accordance with Gelperin and Forsythe's (1975) prediction that oligophagous herbivores would not demonstrate such a type of learning whereas polyphagous ones would. This is because the latter may consume palatable although poisonous plants whereas the finickiness of the oligophagous herbivores prevents such encounters. Presumably, therefore, no selection has taken place for aversion learning ability.

The neural basis of food aversion learning in insects is unknown (Dethier 1982).

Another type of learning to reject a food was observed by Blaney and Winstanley (1982) with *Locusta migratoria*. At the first encounter this insect may palpate and bite before it rejects a nonhost plant. However, the insect seems to associate the sensation at palpation with that at biting, because at subsequent encounters rejection is triggered by palpation alone. Although the result is the same, this process cannot be regarded as aversion learning sensu stricto since no toxic effect can develop, because of the immediate behavioral response. In this case the insect learns to reject an unpalatable food by the sensory input from the mouthparts.

Wasserman (1982) reported that third- to fourth-instar larvae of *Lymantria dispar*, when reared on the least preferred host *Prunus serotina*, showed reduced preference for this plant compared with larvae reared on more acceptable hosts. This may appear as aversion learning; however, Wasserman assumed that this apparent "reversed induction" resulted from the reduced sensitivity of the chemoreceptors to any host when the larvae were reared on the more acceptable hosts.

Many more investigations into such kinds of learning processes are needed before any conclusion can be drawn concerning their incidence among insects and their role in foraging behavior.

Concluding Remarks

The studies on the effects of experience on feeding and egg-laying behavior of phytophagous insects show a variety of phenomena. The common features of these processes are that via chemoreceptors, and in some cases visual receptors, the insect receives very detailed information from the plant; this information is in some way stored in the nervous system, and is used as reference information for the decision to be made by the insect at subsequent encounters with plants. At present the neural basis of these processes is largely unknown, and it would therefore be premature to try to fit them into some contemporary ethological category. This would result in mere terminological speculations which are seldom fruitful.

Since the behavioral changes elicited by experience may affect survival and reproduction, i.e., fitness of the insects, it is logical that the relevant literature is rich in assumptions on their adaptive significances.

The adaptive advantages of aversion learning and habituation to deterrents are almost self-evident. The former prevents the consumption of deleterious quantities of poisonous food whereas the latter enables the insect to utilize food resources that have only "unusual" taste qualities but are otherwise suitable as a food.

As regards feeding preference induction, it has been assumed that it reduces the probability that in a mixed plant stand an insect would frequently change food plant species, which has been shown to decrease the efficiency of food utilization (Schoonhoven and Meerman 1978; Scriber 1979, 1981, 1982; Grabstein and Scriber 1982a, b). However, it is difficult to comprehend the adaptive advantage of the "starving-to-death-at-Lucullian-banquets" phenomenon mentioned earlier.

As regards induced oviposition preferences it has been supposed that they enhance foraging efficiency within resource patches (Prokopy et al. 1982a), or that they reduce the likelihood that an experienced female entering a patch of rare host would stay there and continue searching (Prokopy et al. 1985), or that they result in depositing eggs on more abundant host species, thus enhancing higher juvenile survivorship (Rausher 1980). But here again, one cannot help wondering what advantage may arise from the behavior of a butterfly population whose females lay eggs only on one plant species and ignore several other equally suitable host species present in the same community (Jones and Ives 1979; Rausher 1980).

As a matter of fact, induced feeding and oviposition preferences are nothing other than a further restriction in the overall extreme finickiness of phytophagous insect species at the population and/or individual levels. Therefore, the question of the adaptive value of narrow host plant specialization itself should be answered first. Lots of assumptions have been made in this respect including increased efficiency of food utilization, niche segregation, community structure, species packing, and so on. Unfortunately, none of these assumptions proves to be conclusive enough to reveal the selection factors that have resulted in food specialization. Thus the main question is still unanswered. (For details see the review by Fox and Morrow 1981.)

But are our questions adequate?

It is a general custom in evolutionary biology to ask only "How does this or that character enhance survival?" However, it should be more often asked "Does this or that character jeopardize survival?" Many characters that we try to squeeze into the Procrustean bed of selective advantage must be neutral or at least not deleterious concerning selection pressures to which a given species has been exposed up to the present. I think we should consider more often the ideas propounded by Monod (1970), the great "enfant terrible" of modern biology, and regard Nature not so much as a well-planned factory but also as a giant lottery with many unpredictable and unpurposive events. Our difficulties in really understanding Nature often arise from our deterministic thinking which is an unsurmountable obstacle in finding out what belongs to the "factory" and what to the "lottery" in Nature.

As mentioned above, we may conclude that aversion learning as well as most cases of induced preferences have adaptive significances whereas the ex-

treme cases of rigid induced preferences are nonadaptive. It may simply reflect
the limited flexibility of the insects' neural systems which might even reduce
fitness in certain ecological situations. It can be supposed that such situations
occur only temporarily and locally, and that, therefore, these imperfections
have not been eliminated by selection.

I hope that this short survey has shed some light on what we know about
the changes in the behavior of phytophagous insects elicited by experience and
also on how very little we know about the ecological significance of such changes
and about the underlying mechanisms. This group of questions is a small part
of an area that has been characterized by Dethier (1981) in a most lucid paper
as follows (p 48):

> . . . an area of research that has so few practitioners that almost every one is a
> pioneer. It is an area that guards its secret so jealously that the literature is strewn
> with papers that have become the epitaphs of frustrated and disillusioned workers
> who have subsequently sought fame and fortune in other fields.

References

Aboul-Nasr A, Mansour MH, Salem NY (1981) The phenomenon of induction of pref-
    erence in the cotton bollworm *Heliothis armigera*. Z Pflanzenkr Pflanzenschutz
    88:116–122
Ali M (1976) Studies on the induction of food preference in alfalfa ladybird, *Subcoccinella
    24-punctata* L. (Coleoptera: Coccinellidae). Symp Biol Hung 16:23–28
Barbosa P, Greenblatt J, Withers W, Cranshaw W, Harrington EA (1979) Host-plant
    preferences and their induction in larvae of the gypsy moth, *Lymantria dispar*.
    Entomol Exp Appl 26:180–188
Bernays EA (1983) Antifeedants in crop pest management. In: Whitehead DL, Bowers
    WS (eds) Natural Products for Innovative Pest Management. Pergamon Press, Ox-
    ford, pp 259–271
Blaney WM, Winstanley C (1982) Food selection behaviour in *Locusta migratoria*. In:
    Visser JH, Minks AK (eds) Proc 5th Int Symp Insect-Plant Relationships. Pudoc,
    Wageningen, pp 365–366
Bongers W (1965) External factors in the host plant selection of the Colorado beetle
    Leptinotarsa decemlineata Say. Meded Landbouwhogesch Gent 30:1516–1523
Cassidy MD (1978) Development of an induced food plant preference in the Indian stick
    insect, *Carausius morosus*. Entomol Exp Appl 24:287–293
Chew FS (1980) Foodplant preferences of *Pieris* caterpillars (Lepidoptera). Oecologia
    (Berl) 46:347–353
de Boer G, Hanson FE (1982) Chemical isolation of feeding stimulants and deterrents
    from tomato for the tobacco hornworm. In: Visser JH, Minks AK (eds) Proc 5th
    Int Symp Insect–Plant Relationships. Pudoc, Wageningen, pp 371–372
de Boer G, Hanson FE (1984) Foodplant selection and induction of feeding preference
    among host and non-host plants in larvae of the tobacco hornworm *Manduca sexta*.
    Entomol Exp Appl 35:177–193
Dethier VG (1973) Electrophysiological studies of gustation in lepidopterous larvae. II.
    Taste spectra in relation to food-plant discrimination. J Comp Physiol 82:103–134
Dethier VG (1980) Food-aversion learning in two polyphagous caterpillars, *Diacrisia
    virginica* and *Estigmene congrua*. Physiol Entomol 5:321–325

Dethier VG (1981) Taste and smell in their world and ours. The Chancellor's Lecture Series, 1979–1980. University of Massachusetts at Amherst, pp 48–52

Dethier VG (1982) Mechanism of host-plant recognition. Entomol Exp Appl 31:49–56

Dethier VG, Yost MT (1979) Oligophagy and absence of food-aversion learning in tobacco hornworms *Manduca sexta*. Physiol Entomol 4:125–130

de Wilde J, Sloof R, Bongers W (1960) A comparative study of feeding and oviposition preference in the Colorado beetle *(Leptinotarsa decemlineata Say)*. Meded Landbouwhogesch Gent 25:1340–1346

Flowers RW, Yamamoto RT (1982) Feeding on non-host plants by the tobacco hornworm, *Manduca sexta* (Lepidoptera, Sphingidae). Fla Entomol 65:523–530

Fox LR, Morrow PA (1981) Specialization: species property or local phenomenon? Science 211:887–893

Gelperin A, Forsythe D (1975) Neuroethological studies of learning in mollusks. In: Fentress JC (ed) Simpler Networks and Behavior. Sinauer Associates, Sunderland, MA, pp 239–250

Getzova AB, Lozina-Lozinskii LK (1955) [Role of behavior in the process of insects' adaptation to plant diet]. Zool Zh 34:1066–1078 (In Russian)

Gilbert LE (1975) Ecological consequences of a co-evolved mutualism between butterflies and plants. In: Gilbert LE, Raven PH (eds) Coevolution of Animals and Plants. University of Texas Press, pp 210–240

Gill JS (1972) Studies on insect feeding deterrents with special reference to fruit extracts of the neem tree, *Azadirachta indica* A. Juss. Ph.D. Thesis, University of London

Grabstein EM, Scriber JM (1982a) The relationship between restriction of host plant consumption, and postingestive utilization of biomass and nitrogen in *Hyalophora cecropia*. Entomol Exp Appl 31:202–210

Grabstein EM, Scriber JM (1982b) Host-plant utilization by *Hyalophora cecropia* as affected by prior feeding experience. Entomol Exp Appl 32:262–268

Greenblatt JA, Calvert WH, Barbosa P (1978) Larval feeding preferences and inducibility in the fall webworm, *Hyphantria cunea*. Ann Entomol Soc Am 71:605–606

Hanson FE (1976) Comparative studies on induction of food choice preferences in lepidopterous larvae. Symp Biol Hung 16:71–77

Hanson FE, Dethier VG (1973) Role of gustation and olfaction in food plant discrimination in the tobacco hornworm, *Manduca sexta*. J Insect Physiol 19:1019–1034

Hinde RA (1970) Behavioural habituation. In: Horn G, Hinde RA (eds) Short-Term Changes in Neural Activity and Behaviour. Cambridge University Press, pp 3–40

Hopkins AD (1917) A discussion of C.G. Hewitt's paper on "Insect Behavior." J Econ Entomol 10:92–93

Hubert-Dahl ML (1975) Änderung des Wirtswahlverhaltens dreier Biotypen von *Acyrtosiphon pisum* Harris nach Anzucht auf verschiedenen Wirtspflanzen. Beitr Entomol 25:77–83

Iwao S (1959) Some experiments on the host-plant preference in a phytophagous lady beetle, *Epilachna pustulosa* Kono, with special reference to its individual variation. Insect Ecol 8:10–21

Jermy T, Szentesi À (1978) The role of inhibitory stimuli in the choice of oviposition site by phytophagous insects. Entomol Exp Appl 24:458–471

Jermy T, Bernays EA, Szentesi A (1982) The effect of repeated exposure to feeding deterrents on their acceptability to phytophagous insects. In: Visser JH, Minks AK (eds) Proc 5th Int Symp Insect-Plant Relationships. Pudoc, Wageningen, pp 25–32

Jermy T, Hanson FE, Dethier VG (1968) Induction of specific food preference in lepidopterous larvae. Entomol Exp Appl 11:211–230

Johansson AS (1951) The food plant preference of the larvae of *Pieris brassicae* L. Nor Entomol Tidsskr B 8:187–195

Jones RE, Ives PM (1979) The adaptiveness of searching and host selection behaviour in *Pieris rapae* (L.). Aust J Ecol 4:75–86

Kogan M (1977) The role of chemical factors in insect/plant relationships. In: White D (ed) Proc 5th Int Congr Entomol Washington DC. Entomol Soc Am pp 211–227

Kozhančikov IV (1958) [Biological pecularities of the European species of the genus *Galerucella* and differentiation of biological forms in *Galerucella lineola* L.] Tr Zool Inst Akad Nauk SSSR 24:271–322 (In Russian)

Kuznetzov VI (1952) [The question of adaptation in lepidopterous species to new feeding conditions.] Tr Zool Inst Akad Nauk SSSR 11:166–181 (In Russian)

Ma WC (1972) Dynamics of feeding responses in *Pieris brassicae* Linn. as a function of chemosensory input: a behavioural, ultrastructural and electrophysiological study. Meded Landbouwhogesch Wageningen 72-11:1–162

Mark GA (1982) Induced oviposition preference, periodic environments, and demographic cycles in the bruchid beetle *Callosobruchus maculatus*. Entomol Exp Appl 32:155–160

Marlin NA, Miller RR (1981) Associations to contextual stimuli as a determinant of long-term habituation. J Exp Psych 7:313–333

McFarland D (ed) (1981) The Oxford Companion to Animal Behaviour. Oxford University Press, Oxford

Monod J (1970) Le hasard et la nécessité. Edition du Seuil, Paris

Phillips WM (1977) Modification of feeding "preference" in the flea-beetle, *Haltica lythri* (Coleoptera, Chrysomelidae). Entomol Exp Appl 21:71–80

Prokopy RJ (1977) Attraction of *Rhagoletis* flies (Diptera: Tephritidae) to red spheres of different sizes. Can Entomol 109:593–596

Prokopy JR, Averill AL, Cooley SS, Roitberg CA (1982a) Associative learning in egg-laying site selection by apple maggot flies. Science 218:76–77

Prokopy RJ, Averill AL, Cooley SS, Roitberg CA, Kallet C (1982b) Variation in host acceptance pattern in apple maggot flies. In: Visser JH, Minks AK (eds) Proc 5th Int Symp Insect-Plant Relationship. Pudoc, Wageningen, pp 123–129

Prokopy RJ, Papaj DR, Cooley SS, Kallet C (1986) On the nature of learning in oviposition site acceptance by apple maggot flies. Anim Behav 34:98–107

Rausher MD (1978) Search image for leaf shape in a butterfly. Science 200:1071–1073

Rausher MD (1980) Host abundance, juvenile survival, and oviposition preference in *Battus philenor*. Evolution 34:342–355

Saxena KN (1967) Some factors governing olfactory and gustatory responses of insects. In: Hayashi T (ed) Olfaction and Taste II. Pergamon Press, Oxford, pp 799–819

Saxena KN, Schoonhoven LM (1978) Induction of orientational and feeding preferences in *Manduca sexta* larvae for an artificial diet containing citral. Entomol Exp Appl 23:72–78

Saxena KN, Schoonhoven LM (1982) Induction of orientational and feeding preferences in *Manduca sexta* larvae for different food sources. Entomol Exp Appl 32:173–180

Schoonhoven LM (1969) Sensitivity changes in some insect chemoreceptors and their effect on food selection behaviour. Proc K Ned Akad Wet C 72:491–498

Schoonhoven LM (1977) On the individuality of insect feeding behaviour. Proc K Ned Akad Wet C 80:341–350

Schoonhoven LM, Meerman J (1978) Metabolic cost of changes in diet and neutralization of allelochemics. Entomol Exp Appl 24:689–693

Schweissing FC, Wilde G (1979) Predisposition and nonpreference of the greenbug for certain host plants. Environ Entomol 8:1070–1072

Scriber JM (1979) The effects of sequentially switching foodplants upon biomass and nitrogen utilization by polyphagous and stenophagous *Papilio* larvae. Entomol Exp Appl 25:203–215

Scriber JM (1981) Sequential diets, metabolic cost, and growth of *Spodoptera eridania* feeding upon dill, lima bean and cabbage. Oecologia (Berl) 51:175–180

Scriber JM (1982) The behaviour and nutritional physiology of southern armyworm larvae as a function of plant species consumed in earlier instars. Entomol Exp Appl 31:359–369

Smith MA, Cornell HV (1979) Hopkins host-selection in *Nasonia vitripennis* and its implications for sympatric speciation. Anim Behav 27:365–370

Städler E, Hanson FE (1976) Influence of induction of host preference on chemoreception of *Manduca sexta:* behavioral and electrophysiological studies. Symp Biol Hung 16:267–273

Städler E, Hanson FE (1978) Food discrimination and induction of preference for artificial diets in the tobacco hornworm, *Manduca sexta*. Physiol Entomol 3:121–133

Stanton ML (1979) The role of chemotactile stimulation in the oviposition preferences of *Colias* butterflies. Oecologia (Berl) 39:79–81

Strebl O (1928) Biologische Studien an einheimischen Collembolen. II. Ernährung and Geschmacksinn bei *Hypogastrura purpurascens* (Lubb) (Apter., Coll.). Z Wiss Insektenbiol 23:135–143

Stride GO, Straatman R (1962) The host plant relationship of an Australian swallowtail *Papilio aegeus,* and its significance in the evolution of host plant selection. Proc Linn Soc NSW 87:69–78

Szentesi À, Bernays EA (1984) A study of behavioural habituation to a feeding deterrent in nymphs of *Schistocerca gregaria*. Physiol Entomol 9:329–340

Tabashnik BE, Wheelock H, Rainbolt JD, Watt WB (1981) Individual variation in oviposition preference in the butterfly, *Colias eurytheme*. Oecologia (Berl) 50:225–230

Thompson RF, Spencer WA (1966) Habituation: a model phenomenon for the study of neuronal substrates of behaviour. Psychol Rev 73:16–43

Ting AY (1970) The induction of feeding preference in the butterfly *Chlosyne lacinia*. Unpublished thesis, University of Texas, Austin

Traynier RMM (1979) Long-term changes in the oviposition behaviour of the cabbage butterfly, *Pieris rapae,* induced by contact with plants. Physiol Entomol 4:87–96

Wasserman SS (1982) Gypsy moth (*Lymantria dispar*) induced feeding preferences as a bioassay for phenetic similarity among host plants. In: Visser JH, Minks AK (eds) Proc 5th Int Symp Insect-Plant Relationships. Pudoc, Wageningen, pp 261–267

Wiklund C (1973) Host plant suitability and the mechanism of host selection in larvae of *Papilio machaon*. Entomol Exp Appl 16:232–242

Wiklund C (1974) Oviposition preferences in *Papilio machaon* in relation to the host plants of the larvae. Entomol Exp Appl 17:189–198

Wiklund C (1982) Generalist versus specialist utilization of host plants among butterflies. In: Visser JH, Minks AK (eds) Proc 5th Int Symp Insect-Plant Relationships. Pudoc, Wageningen, pp 181–191

Wiseman BR, McMillian W W (1980) Feeding preferences of *Heliothis zea* larvae preconditioned to several host crops. J Ga Entomol Soc 15:449–453

Yamamoto RT (1974) Induction of hostplant specificity in the tobacco hornworm, *Manduca sexta*. J Insect Physiol 20:641–650

Chapter 10

The Evolution of Deterrent Responses in Plant-Feeding Insects

ELIZABETH BERNAYS AND REG CHAPMAN*

A plant deterrent may be defined as "a chemical which inhibits feeding or oviposition when present in a place where insects would, in its absence, feed or oviposit" (Dethier et al. 1960). In general, we believe that behavioral deterrence caused by such chemicals in plants plays a major role in host selection (Dethier 1954; Jermy 1966; Bernays and Chapman 1977), and the chemicals effectively protect most plants from most insects. All plants have a complex profile of secondary compounds ranging in number from a few to hundreds and there are an estimated 100,000–400,000 different secondary compounds in terrestrial plant species (Swain 1977). Most that have been investigated are deterrent to one or other of the insect species tested. The more restricted the insect's host range, the more compounds that are found deterrent and the lower the threshold for rejection of such deterrents (Jermy 1983).

It is widely believed that the primary defenses of plants are those that act post-ingestively and that behavioral rejection by insects developed as an adaptive response to this toxicity. Many recent papers demonstrating the deterrent effects of various plant secondary compounds take this adaptive link for granted (Rhoades 1983; Crawley 1983; Krischik and Denno 1983) and in a recent review Berenbaum (1986) states that it is practically beyond questioning that compounds causing behavioral deterrence will be associated with disadvantages if ingested. This view is taken particularly with respect to insects because their evolutionary adaptability is often considered to be such that harmless resources are unlikely to remain unutilized, and that the possession of a harmless compound simply causing behavioral deterrence is ultimately no protection against insect herbivory.

The sequence of events leading to the evolution of the deterrent response is thus assumed to be:

1. The plant evolves enzymic pathways to produce and maintain a novel compound.

*Division of Biological Control and Department of Entomological Sciences, University of California, Berkeley, California 94720, U.S.A.

2. If the compound has toxic properties but no deterrent effect, the animals utilizing such a plant as a food source could survive by (a) evolving suitable physiological tolerance mechanisms or (b) evolving behavioral rejection.

With respect to (a), numerous examples exist of how particular insect species are adapted physiologically to diverse biologically active compounds (Rosenthal and Janzen 1979), and no plant is without its guild of adapted insect herbivores (Lawton 1978). With respect to (b), deterrent responses to substances such as alkaloids, which are well-known to be toxic to vertebrates, are extremely common. Evolved resistance of insects to man-made insecticides can include a behavioral avoidance and may thus parallel evolved deterrent responses of insects in relation to plant toxins (Pluthero and Singh 1984).

Given that rejection of plants governed by deterrent compounds is so important among insects the evolution of deterrent responses to harmful plant products must logically have occurred on innumerable occasions and one would expect a strong correlation between behavioral deterrence and adverse postingestive effects.

The Deterrence/Toxicity Relationship

A critical evaluation of the literature reveals that an unequivocal link between deterrence and antibiosis or toxicity has rarely been demonstrated though examples do exist. Waldbauer (1962), for example, showed that when chemoreceptors were removed from caterpillars of *Manduca sexta* the larvae fed and grew on some plants that were normally avoided, but their ability to utilize the ingested nonhost was reduced compared with their ability to utilize the usual hosts. The implication is that the secondary compounds in the nonhosts, although no longer perceived as deterrent, reduced the efficiency of digestion and assimilation.

Specific deterrent secondary compounds have also been shown to be toxic following ingestion. For example, Erickson and Feeny (1974) and Blau et al. (1978) showed that glucosinolates were both deterrent and toxic for *Papilio polyxenes* caterpillars, which normally feed on Umbelliferae. High concentrations of tannic acid are deterrent to grass-feeding grasshoppers, and if ingested this material causes gut lesions (Bernays et al. 1980).

Numerous other experiments purport to show reduced digestion, assimilation, or other "toxic" effects, but most do not clearly separate a deterrent response leading to reduced food intake from antibiotic effects. This distinction is important because even a small reduction in the intake of a normal food without deterrents may have effects on digestive efficiency and assimilation (Schroeder 1976). Thus deterrence can lead to false measures of antibiotic effects and must be specifically taken into account.

There are some instances where the absence of a link between deterrence and toxicity has been demonstrated. Harley and Thorsteinson (1967) showed a poor correlation between deterrence and toxicity of a range of different secondary compounds with *Melanoplus bivittatus,* and in experiments with *Locusta migratoria* some deterrents had no measurable effect on digestive parameters or on overall growth. Boys (1981) studied the effects of ingestion of the alkaloid gramine on *Locusta migratoria* and was unable to demonstrate any deleterious effect on growth or fecundity at naturally occurring concentrations, although it was strongly deterrent in the natural concentration range. Various condensed tannins and cyanogenic glycosides deter grasshoppers from feeding, but when added to the normal food plant and presented without choice no deleterious effects could be measured (Bernays et al. 1981; Bernays 1983). Hsiao and Fraenkel (1968), studying *Leptinotarsa decemlineata,* showed that some plants other than the normal solanaceous hosts supported good growth and reproduction. Thus *Asclepias* and *Lactuca* were suitable food plants although normally rejected as a result of unidentified deterrents. Furthermore, a laboratory strain of these insects has rapidly become behaviorally tolerant of some *Solanum* spp. that are relatively deterrent to field populations (Hsiao 1982). All these experiments were possible only because the insects had no choice but to eat the deterrents. Rejection of the deterrent food would have occurred if more palatable food had been available.

Strong deterrents cannot easily be tested for their oral toxicity, and experimental methods must be developed that circumvent this problem. One method for dealing with moderately strong deterrents is to match reduced amounts eaten by insects on a deterrent diet with similar limited quantities of food in the controls. Regression analyses of various growth indices against amount eaten may then give estimates of effects over and above the effect of lowered consumption. This method has been used by Usher and Feeny (1983) in testing the effects of cucurbitacins, cardenolides, and alkaloids from certain crucifers on larvae of *Pieris rapae.* The only effects they found were increases in growth. Another approach is to bypass the chemoreceptors altogether. For example, a cannula may be used to place materials directly into the gut, or the insects may be dosed with capsules that release their contents after ingestion. These methods have been used with acridids (Szentesi and Bernays 1984). Cottee (1984), also using these methods, showed that of seven different deterrents some had measurable deleterious effects and others did not.

Among Lepidoptera, rejection of many plant species by ovipositing females is often not an indication of plant unsuitability as food for larvae. Wiklund (1982), for example, found that female *Anthocharis cardamines* rejected several common field plants, although these will support larval growth as well as the normal host.

Proof of the absence of antibiotic effects, however, will also require extensive investigation of the long-term effects on fecundity. Experiments with larvae alone can be misleading. For example, Brattsten (1983) showed that the monoterpene pulegone from mint increased the wet weight of *Spodoptera eridania* pupae when added to the larval diet but the emerging adults laid fewer eggs compared with controls.

Considerably more work is needed, therefore, before any generalizations can be made, but the suggestion so far is that the relationship between deterrence and toxicity may not be as tight as is commonly presumed. If experimental work continues to show that deterrence is extremely widespread but not matched by any toxic effects, the assumption that secondary compounds are primarily antiherbivore defenses must be reassessed. It has recently been questioned on other grounds (Jermy 1984), but a poor deterrence–toxicity relationship will give grounds for seriously doubting many "defensive" theories.

Evolution of the Deterrent Response

If deterrent responses arise as a consequence of the toxicity of plant secondary compounds, at least on some occasions, how could selection operate to favor the development of this behavior? Although this is a critical step, we find little consideration of this point in the literature (Gould 1984). The following comments are no more than speculation.

We assume that a deterrent response to a particular toxic compound could evolve only if some feature of the insect's biology continued to bring it habitually into contact with the new compound. This could happen in several different ways. It is possible that deterrent responses might have evolved among those insects that exhibited some degree of physiological tolerance of the new compound from the outset. One can envisage that, whereas selection tended to enhance tolerance in one segment of the population, it may have favored a deterrent response in another segment provided they had the capacity to develop on an alternative host.

In polyphagous or very mobile species, effective deterrence may be expected to develop if the plant concerned is very common. For example, in the case of *Schistocerca gregaria*, the desert locust, exceptional behavioral sensitivity has developed to azadirachtin, a potent toxin from the neem tree *Azadirachta indica*. The tree and the insect probably share a geographical origin and, in the very arid environment, the tree may have comprised a major part of the available vegetation so that interactions between plant and insect were a common occurrence. On a smaller scale, selection for a deterrent response may occur where insects feeding on a clump of one plant species frequently encounter a phenotype with a new toxic principle.

Among species where the ovipositing female selects the host, continued egg-laying on a plant on which larval survival was poor would favor the development of oviposition deterrent behavior.

Insects may evolve avoidance of plants through food aversion learning if the plant causing toxicity is common enough. Food aversion learning has been demonstrated in two polyphagous species (Dethier 1980a) but not in two oligophagous ones (Dethier and Yost 1979); if it is a regular phenomenon it may be that deterrent behavior develops by the gradual genetic fixation of the aversive responses as suggested for mollusks by Gelperin and Forsythe (1976) and discussed by Mitter and Futuyma (1983).

Changes Due to Experience

Habituation to deterrents (as opposed to increased ingestion through "hunger") is an indication of the relative insignificance of some deterrents physiologically. Several species of phytophagous insects have been shown to habituate to some deterrent plant secondary compounds (see Jermy, this volume). For example, Jermy et al. (1982) showed that the larvae of *Mamestra brassicae* and *Pieris brassicae* habituated to strychnine and quinine in the diet so that although consumption by naive larvae was reduced by these deterrents, experienced larvae ate amounts similar to those eaten by control larvae, and reached similar final weights, with no increase in mortality. The same was true for the locust, *Schistocerca gregaria*, fed on sorghum leaves to which nicotine hydrogen tartrate was added. Insects that had nicotine placed directly in the gut via a cannula or in gelatin capsules over a 4-day period subsequently ate far more deterrent-treated food than did naive control insects (Szentesi and Bernays 1984). The test insects developed as quickly as controls and reached similar final weights.

Intraspecific Variation in Host Plant Use

Observations of the natural feeding habits of insects sometimes indicate a poor relationship between deterrence and long-term unsuitability of food. Futuyma et al. (1984) describe an example of the differential acceptance of a host, maple, by two genotypes of larvae of the moth *Alsophila pometaria*. Despite the fact that both perform equally well on this host, one genotype disperses more readily from it, presumably because it is deterred by certain chemical(s), which are not deleterious when ingested. Furthermore, many insect herbivore species show local differences in host plant preference that do not necessarily reflect differences in suitability of the plants for growth, but are apparently adaptations to other factors such as local patterns of predation, relative host abundance, and plant phenology (Fox and Morrow 1981).

Rapid behavioral changes leading insects readily to accept plants that were initially less acceptable can be induced in the laboratory. The larvae of many species of Lepidoptera develop a preference for the plants upon which they have been feeding for 48 hr or more, although other plant species may allow equally good growth and fecundity (Hanson 1983). Such effects suggest some positive value in maintaining the use of single host plant species through larval life. Host shifts, involving genetic change, onto originally nonpreferred plants (including cultivated crops) have occurred in the field in the last 50 years (Futuyma 1983a; Bush and Hoy 1984). For example, among populations of the apple maggot fly, *Rhagoletis pomonella*, variations in host use exist that appear to depend entirely on behavioral choices without accompanying variation in suitability for larval development (Prokopy et al. 1982). Such host changes may involve unrelated and chemically dissimilar species of plants and indicate either that physiological adaptations occur with equal rapidity to the behavioral changes or that they are not necessary. Dethier (1954, 1970) considered that behavioral lability was greater than that of any physiological parameters, implying that deterrence was much more widespread than any accompanying toxicity. Wasserman and Futuyma (1981) make a similar suggestion and it does seem that rapid changes of preference and host range indicate a much greater behavioral lability than would be expected in a situation having the constraints of a tight deterrence–toxicity relationship.

Chemical Diversity and Insect Host Range

If the deterrent response is not primarily an adaptation to avoid toxic compounds, what is its value to the insect herbivore? The secondary compounds may be considered less as defenses and more as signals of nonhosts and hosts, and cues for learning in relation to food quality. Since deterrents play a dominant role in determining host range, the question becomes: what is the value of such a restricted host range? Perhaps, as suggested for vertebrate predators (Pyke et al. 1977), specialization enhances the efficiency of host location, providing the host remains relatively abundant (Futuyma 1983b).

Many insect herbivores exhibit extreme habitat specialization. Stem borers, leaf miners, and gall formers are groups with a high proportion of monophages (Price 1983) and one may suppose that structural as well as physiological modifications may require that the insects be very selective. Similarly, specialization of the mouthparts is another constraint on the host range of some groups. Among grass-feeding grasshoppers, the molar region of the mandibles is adapted for grinding and such insects may shear the ab- and adaxial surfaces of grass leaves apart. A resultant digestive superiority on grasses would make it advantageous to stay on grasses (Boys 1981; Bernays and Barbehenn 1986).

The small size of insects means that specialized mechanisms are required for mate-finding from any distance. A growing number of examples is being found where plant-specific cues are used to locate the area where the mate may be found. This may be particularly important in holometabolous insects whose larvae and adults occupy different habitats (Labeyrie 1978). Many insects use host plant secondary compounds as precursors of specific pheromones associated with mate-finding and other behaviors: they must feed on the appropriate host to obtain the compounds. The best known examples of this requirement are coniferous bark beetles (Wood 1982), but other examples are cited by Rodriguez and Levin (1976); see also Schneider (this volume).

Many insect species currently have a tight phenological relationship with particular plants, ensuring that high nutrient food is available for young larvae, whose protein needs are most acute. Examples are given in Strong et al. (1984) that demonstrate that seasonal phenology in relation to the host plant is a dominant influence in populations of many phytophagous species. Phenological matching may also involve larval growth, which ensures appropriate conditions for the adults, such as flowering of suitable plants for pollen or nectar feeding. Many other complexities as guild members have been described that involve dependance on particular plants (Rathke 1976; Price et al. 1980).

Lawton (1978) suggests that "avoidance of predators and parasitoids is a major force driving niche diversification," and it must be noted that host-specific crypsis is extremely widespread. Schultz (1983) gives many examples with tree-feeding caterpillars and shows that such crypsis is of vital importance in survival. Another common example is among grasshoppers, where over 50% of African species show extreme grass-specific crypsis (Bernays and Barbehenn 1986). Sequestering of plant compounds broadly toxic to vertebrates is another surprisingly common strategy among herbivorous insects (Duffey 1980; Pasteels and Rowell-Rahier 1983). The value of this is enhanced by the greater overall physiological sensitivity of vertebrates to alkaloids and other toxins compared with that of insects (van Emden 1978; Bernays 1982). Both these strategies utilize characteristics of the host plant for protection, and maintaining rejection responses of inappropriate plants and positive attraction to specific host plants is essential to the value of the protective devices. A variety of additional possible reasons for "ecological monophagy" are suggested by Futuyma (1983a).

An analysis of Berenbaum's data (1981) on furanocoumarins in the Umbelliferae shows that increasing number and complexity of chemicals are associated with more species of both specialist and generalist insect herbivores, although the numbers involved are small. This would suggest some positive effect for insects in spite of the current conventional idea that such an array of biologically active secondary compounds is an indication of specialized defensive chemistry.

From this brief survey it appears that the positive values of host restriction may provide a major driving force for the development of specialized feeding habits and behavioral sensitivity to deterrents. If chemical diversification has preceded insect diversification (Futuyma 1983a; Jermy 1984), insects may positively utilize the diversity to reinforce specificity. Evidence is mounting that secondary plant compounds may serve insects in different ways although most theories on evolutionary mechanisms of plant deterrence to insects continue to focus on specific anti-insect defenses. Can a study of the chemoreceptors shed any light on the subject?

The Receptor System

Deterrence is a behavioral phenomenon, so the evolutionary development of deterrent behavior cannot be separated from the evolution of the sensory receptors involved in the perception of deterrents. The question of the evolution of receptor sensitivity to plant secondary compounds was addressed by Dethier in 1980 (Dethier 1980b).

Among phytophagous insects relevant neurophysiological studies are largely restricted to the Acridoidea and larval Lepidoptera. The only extensive work on the chemosensory system of Acridoidea is that of Blaney (1974, 1975, 1980) on *Locusta migratoria* and *Schistocerca gregaria*. Blaney (1974) found that every chemical tested produced activity in several (usually more than two) receptor cells in any sensillum, irrespective of whether that substance produced acceptance or rejection behavior. Moreover, to judge from the amplitude of the spikes, each cell was stimulated by a range of different substances that might also vary in their behavioral quality (Blaney 1975); each cell has a broad spectrum of response.

Blaney's subsequent analyses of the relationship between sensory input and behavior (Winstanley and Blaney 1978; Blaney 1980; Blaney and Winstanley 1980), are based on the assumption that the interpretations by the insect of the inputs from the different cells, or different sensilla, are not distinguished in any qualitative manner. If this is correct, central differentiation of the inputs from cells as stimulating or inhibiting, as occurs in Lepidoptera for instance, followed as a later step.

However, Blaney (1980) concludes that his data on *Schistocerca gregaria* indicate that this species has a receptor for azadirachtin whose input leads specifically to rejection. This may be a specific exception to the more generalized nature of other cells, but only further analysis can resolve this problem.

It is known that in many contact chemoreceptors of endopterygote phytophagous insects at least one cell is particularly sensitive to substances that induce a deterrent response and it has been labelled the deterrent receptor. It has been particularly well studied in larval Lepidoptera. In almost every case where a range of compounds has been tested, the cell is sensitive to chemicals in more than one chemical class. In those cases where a lack of sensitivity to a class of compounds is apparent, only a few chemicals have been tested so it is not yet possible to state categorically that these receptors are limited in the classes of chemicals to which they respond. In no case so far investigated, however, does the cell respond to all chemical classes tested or even to all the chemicals within a class. For example, the deterrent receptor of *Manduca sexta* responds to the flavonoid phlorizin, but not to quercetin or rutin; and it responds to salicin, but not to arbutin (Schoonhoven 1981). The pattern differs from insect to insect (Schoonhoven 1981) even in closely related species. Among the Sphingidae, for example, there is evidence of an electrophysiological response to some alkaloids, phenolic glycosides, and glucosinolates in *Manduca sexta*, but the same compounds are not stimulating to *Sphinx ligustri*. Within the genus *Papilio* the responses of three species to sinigrin, quercitrin, and morin vary from species to species (Dethier and Kuch 1971; Dethier 1973), and some variation is also evident within the genus *Yponomeuta* (van Drongelen 1979).

It is significant that the deterrent cells have commonly been shown to respond to compounds that they cannot have experienced in their recent evolution. Azadirachtin is a good example. This triterpenoid is obtained from the neem tree, *Azadirachta indica*. Although this tree has recently been introduced into many tropical areas it originated in India; yet the deterrent cell of a number of Palearctic and Neotropical species is sensitive to the compound (Schoonhoven 1981; Simmonds and Blaney 1984). It is not known if the cell is sensitive to other triterpenoids. A similar argument applies to the alkaloid strychnine which is derived from tropical trees of the genus *Strychnos*, although in this case most of the insects examined are known to possess a cell that is also sensitive to other alkaloids.

In these cases the receptor cannot have evolved a special sensitivity to these compounds since it had not previously been exposed to them. The likelihood is, however, that it did already have a sensitivity for the appropriate classes of compound.

We can find no evidence of adaptive *development* of sensitivity to a specific deterrent compound, but there is evidence of a *loss* of sensitivity. Van Drongelen (1979) has examined the sensitivity patterns of the receptors in a range of closely related *Yponomeuta* species living on different host plants. These are believed to be derived from an ancestor feeding on Celastraceae (Gerrits-Heybroek et al. 1978). The species that currently live on Celastraceae have a deterrent cell sensitive to phlorizin and salicin, compounds that are present in high concentrations in *Malus* and *Salix*, respectively. Sensitivity to phlorizin is also present in other *Yponomeuta* species, but is absent from *Y. malinellus*, which feeds on apple. In *Y. rorellus*, feeding on *Salix*, the deterrent cell in the lateral galeal sensillum is insensitive to salicin, but that in the medial sensillum is sensitive to it; in other species both receptors are sensitive to salicin. It seems that in these cases the insects that have moved on to new host plants have lost, or are in the process of losing, their sensitivity to compounds that are specific to these hosts. Van Drongelen (1979) argues that the larval switch must almost certainly have followed a change in adult behavior since larval distribution is largely governed by adult oviposition. The sensitivity change is thus an adaptive one occurring *after* the behavioral change of the adult.

A less clear-cut example of a possible loss of sensitivity by a deterrent cell is shown in the responses of the cells to glucosinolates in the larvae of 12 different species of Lepidoptera. Five polyphagous species possess a deterrent cell sensitive to glucosinolates, although some of them are known to feed on crucifers. A similar sensitivity occurs in two of three monophagous species, but apparently not in *Danaus plexippus;* none of these will feed on crucifers. Among oligophagous species, *Papilio polyxenes*, feeding principally on Umbelliferae, and *Manduca sexta*, feeding on Solanaceae, have a deterrent receptor sensitive to glucosinolates, but in *Pieris brassicae* this sensitivity is lacking; instead a separate cell is now sensitive to the glucosinolates and this signals acceptance rather than rejection (Ma 1972; Schoonhoven 1973). So in this series, too, it may be argued that there has been an evolutionary loss of deterrent cell sensitivity in the species specializing on glucosinolate-containing plants.

Except for some particular situations outlined on page 162, encounters by an insect with the multiplicity of plant secondary compounds in nonhost plants will generally be very intermittent and it is hard to conceive that sufficient selection pressure would be maintained for sensory modification to occur. For this reason, we believe that the development of sensitivity to specific deterrent compounds may be a relatively uncommon phenomenon.

Apart from stimulating sensory cells in the normal manner, it is known that deterrent compounds may inhibit the activity of some cells that normally stimulate feeding (Dethier 1982; Schoonhoven 1982; Mitchell and Sutcliffe 1984). This type of interaction is not peculiar to deterrents (Schoonhoven 1973) and the activity of the deterrent cell may itself be depressed by sugar. Insufficient critical information is available on this type of interaction to permit evaluation of its significance in the context of the evolution of deterrent behavior.

Other compounds that deter feeding appear to have a damaging effect on the cells in a sensillum. Quinine is the most widely investigated of such substances. Dethier (1980b) observed that all four chemosensitive cells in the medial and lateral sensilla of *D. plexippus* fired in response to stimulation by this alkaloid. Mitchell and Sutcliffe (1984) obtained very variable responses in *Entomoscelis americana* and Blaney and Simmonds (1983) observed bursting responses in the receptor of *Spodoptera littoralis* and *Heliothis zea*. Subsequently the response of the sensillum to salt was reduced. Similar effects of damage to the chemoreceptor neurons by quinine have been reported in other caterpillars and in flies (Dethier 1982). This effect is not universal because sometimes normal responses occur (Mitchell and Sutcliffe 1984) or there is no response, as in *Heliothis virescens* (Blaney and Simmonds 1983). Nevertheless it is likely that quinine commonly leads to a deterrent response by an insect by virtue of its interference with the input from receptors other than deterrent receptors. Mitchell and Harrison (1985) report a similar phenomenon in *Leptinotarsa decemlineata* when stimulated by the glycoalkaloids solanine, chaconine, and tomatine and the saponin digitonin, and Ma (1977) has documented the inhibition of the response to sugar following stimulation by the sesquiterpene dialdehyde warburganal.

The fact that these compounds act by damaging the receptors suggests that they may, indeed, have widespread antibiotic effects. Although habituation to quinine has been found (Jermy et al. 1982) and critical data are lacking, we might expect a closer link between deterrence and toxicity with such compounds. Their effect apparently does not entail the evolution of a deterrent receptor (Mitchell and Harrison 1985).

If present-day Orthoptera can be regarded as possessing a more generalized (primitive) nervous system than is found in endopterygote insects (Chapman 1982), it appears that contact chemoreceptors have evolved through a stage of broad-spectrum sensitivity, to the more restricted ranges found in most cells of Endopterygota. As a result of the wide sensitivity of the cell many compounds that are novel to the insect are interpreted as being deterrent. We emphasize that deterrent behavior induced by such compounds is simply a consequence of the sensitivity of the deterrent cell; it is in no way a reflection of toxic effects that the compound may or may not have on the insect. Dethier (1980b) has previously concluded that the deterrent receptors of herbivorous insects have evolved through a stage of sensitivity to a wide variety of compounds rather than developing specific responses to particular compounds.

In the context of the whole plant, the input from the deterrent cell is affected by interactions between different plant constituents. Nevertheless, as Dethier (1982) points out, acceptance or rejection depends on a balance between stimulating and inhibiting inputs in which the activity of the deterrent cell plays a part. Given the broad spectrum of response of this cell, it should come as no surprise that novel foods are commonly rejected by phytophagous insects. Indeed, this is just what we might expect. Rejection in such cases is a consequence of the sensory perception of the insect; it does not reflect any ecological adaptation.

Conclusions

We conclude that there are reasons for supposing that the deterrent response is commonly not an adaptive response to the toxicity of plant secondary compounds. We believe that such behavior may often be simply a consequence of the broad range of sensitivity of the sensilla in an environment with a great diversity of compounds, and it leads the insect to avoid unusual situations. Deterrent responses consequently contribute to host plant specificity and we have argued that there are good reasons for host specificity that are unrelated to plant toxicity.

This is not to say that links between deterrence and toxicity do not sometimes occur. They undoubtedly do, and in some of these it is possible that deterrence is a direct consequence of toxicity. Our purpose here, however, is to emphasize that the link is not universal, and to plead for a more critical approach to the subject.

References

Berenbaum M (1981) Patterns of furanocoumarin distribution and insect herbivory in the Umbelliferae: plant chemistry and community structure. Ecology 62:1254–1266

Berenbaum M (1986) Post-ingestive effects of phytochemicals on insects: on Paracelsus and plant products. In: Miller TA, Miller J (eds) Insect-Plant Interactions. Springer-Verlag, New York

Bernays EA (1982) The insect on the plant: A closer look. In: Visser JH, Minks AK (eds) Proc 5th Int Symp Insect-Plant Relationships. Pudoc, Wageningen, pp 3–17

Bernays EA (1983) Nitrogen in defense against insects. In: Lee JA, McNeill S, Rorison IH (eds) Nitrogen as an Ecological Factor. Blackwell, Oxford, pp 321–344

Bernays EA, Barbehenn R (1986) Nutritional ecology of grass foliage-chewing insects. In: Slansky F, Rodriguez JG (eds) Nutritional Ecology of Arthropods. Wiley, New York (in press)

Bernays EA, Chapman RF (1977) Deterrent chemicals as a basis of oligophagy in *Locusta migratoria*. Ecol Entomol 2:1–18

Bernays EA, Chamberlain D, McCarthy P (1980) The differential effects of ingested tannic acid on different species of Acridoidea. Entomol Exp Appl 28:158–166

Bernays EA, Chamberlain D, Leather E (1981) Tolerance of acridids to ingested condensed tannin. J Chem Ecol 7:247–256

Blaney WM (1974) Electrophysiological responses of the terminal sensilla on the maxillary palps of *Locusta migratoria* (L.) to some electrolytes and non-electrolytes. J Exp Biol 60:275–293

Blaney WM (1975) Behavioural and electrophysiological studies of taste discrimination by the maxillary palps of larvae of *Locusta migratoria* (L.). J Exp Biol 62:555–569

Blaney WM (1980) Chemoreception and food selection by locusts. In: van der Starre H (ed) Olfaction and Taste VII. Informational Retrieval, London, pp 127–130

Blaney WM, Simmonds MSJ (1983) Electrophysiological activity in insects in response to antifeedants. Unpublished report, Birkbeck College, University of London

Blaney WM, Winstanley C (1980) Chemosensory mechanisms of locusts in relation to feeding: the role of some secondary plant compounds. In: Insect Neurobiology and Pesticide Action (Neurotox 79). Society of Chemical Industry, London, pp 383–389

Blau PA, Feeny P, Contardo L, Robson DS (1978) Allyglucosinolate and herbivorous caterpillars: a contrast in toxicity and tolerance. Science 200:1296–1298

Boys H (1981) Food selection by some graminivorous Acrididae. D Phil Thesis, University of Oxford

Brattsten L (1983) Cytochrome P-450 involvement in the interactions between plant terpenes and insect herbivores. In: Hedin PA (ed) ACS Symposium Plant Resistance to Insects. American Chemical Society, pp 173–195

Bush GL, Hoy MA (1984) Evolutionary processes in insects. In: Huffaker CB, Rabb RL (eds) Ecological Entomology. John Wiley, New York, pp 247–278

Chapman RF (1982) Chemoreception. The significance of sensillum numbers. Adv Insect Physiol 16:247–356

Cottee P (1984) A physiological investigation into the role of secondary plant compounds as feeding deterrents. PhD thesis, University of Aberdeen

Crawley MJ (1983) Herbivory. Blackwells, Oxford

Dethier VG (1954) Evolution of feeding preferences in phytophagous insects. Evolution 8:33–54

Dethier VG (1970) Chemical interactions between plants and insects. In: Sondheimer E, Simeone JB (eds) Chemical Ecology. Academic Press, New York, pp 83–102

Dethier VG (1973) Electrophysiological studies of gustation in lepidopterous larvae II. Taste spectra in relation to food-plant discrimination. J Comp Physiol 82:103–134

Dethier VG (1980a) Food aversion learning in two polyphagous caterpillars, *Diacrisia virginica* and *Estigmene congrua*. Physiol Entomol 5:321–325

Dethier VG (1980b) Evolution of receptor sensitivity to secondary plant substances with special reference to deterrents. Am Natur 115:45–66

Dethier VG (1982) Mechanism of host-plant recognition. Entomol Exp Appl 31:49–56

Dethier VG, Kuch JH (1971) Electrophysiological studies of gustation in lepidopterous larvae. I. Comparative sensitivity to sugars, amino acids and glycosides. Z Vgl Physiol 72:343–363

Dethier VG, Yost MT (1979) Oligophagy and absence of food aversion learning in tobacco hornworms *Manduca sexta*. Physiol Entomol 4:125–130

Dethier VG, Barton Browne L, Smith NS (1960) The designation of chemicals in terms of the responses they elicit from insects. J Econ Entomol 53:134–136

Duffey SS (1980) Sequestration of plant natural products by insects. Annu Rev Entomol 25:447–477

Erickson JM, Feeny P (1974) Sinigrin: a chemical barrier to the black swallowtail butterfly, *Papilio polyxenes*. Ecology 55:103–111

Fox LR, Morrow PA (1981) Specialization: species property or local phenomenon? Science 211:887–893

Futuyma DJ (1983a) Evolutionary interactions among herbivorous insects and plants. In: Futuyma DJ, Slatkin M (eds) Coevolution. Sinauer, Sunderland, MA, pp 207–231

Futuyma DJ (1983b) Evolutionary aspects of host selection. In: Ahmad S (ed) Herbivorous Insects. Academic Press, New York, pp 227–244

Futuyma DJ, Cort RP, Noordwijk I (1984) Adaptation to host plants in the fall cankerworm (*Alsophila pometaria*) and its bearing on the evolution of host affiliation in phytophagous insects. Am Natur 123:287–296

Gelperin A, Forsythe D (1976) Neuroethological studies of learning in mollusks. In: Fentress JC (ed) Simpler Networks and Behavior. Sinauer, Sunderland, MA, pp 239–250

Gerrits-Heybroek EM, Herrebout WM, Ulenberg SA, Wiebes JT (1978) Host plant preference of five species of small ermine moths (Lepidoptera: Yponomeutidae). Entomol Exp Appl 24:160–168

Gould F (1984) Role of behavior in the evolution of insect adaptation to insecticides and resistant host plants. Bull Entomol Soc Am 30(4):34–41

Hanson F (1983) The behavioral and neurophysiological basis of foodplant selection by lepidopterous larvae. In: Ahmad S (ed) Herbivorous Insects: Host Seeking Behavior and Mechanisms. Academic Press, New York, pp 3–23

Harley KLS, Thorsteinson AJ (1967) The influence of plant chemicals on the feeding behavior, development and survival of the two-striped grasshopper, *Melanoplus bivittatus* Say, Acrididae, Orthoptera. Can J Zool 45:305–319

Hsiao TH (1982) Geographic variation and host plant adaptation of the Colorado potato beetle. In: Visser JH, Minks AK (eds) Proc 5th Int Symp Insect-Plant Relationships. Pudoc, Wageningen, pp 315–324

Hsiao TH, Fraenkel G (1968) Selection and specificity of the Colorado potato beetle for solanaceous and non-solanaceous plants. Ann Entomol Soc Am 61:493–503

Jermy T (1966) Feeding inhibitors and food preference in chewing phytophagous insects. Entomol Exp Appl 9:1–12

Jermy T (1983) Multiplicity of insect antifeedants in plants. In: Whitehead DL, Bowers WS (eds) Natural Products for Innovative Pest Management. Pergamon Press, New York, pp 223–236

Jermy T (1984) Evolution of insect/host plant relationships. Am Natur 124:609–630

Jermy T, Bernays EA, Szentesi À (1982) The effect of repeated exposure to feeding deterrents on their acceptability to phytophagous insects. In: Visser JH, Minks AK (eds) Proc 5th Int Symp Insect Plant Relationships. Pudoc, Wageningen, pp 25–32

Krischik VA, Denno RF (1983) Individual, population, and geographic patterns in plant defense. In: Denno RF, McClure MS (eds) Variable Plants and Herbivores in Natural and Managed Systems. Academic Press, New York, pp 463–512

Labeyrie V (1978) Reproduction of insects and coevolution of insects and plants. Entomol Exp Appl 24:496–504

Lawton JH (1978) Host-plant influences on insect diversity: the effects of space and time. In: Mound LA, Waloff N (eds) Diversity of Insect Faunas. Blackwell, Oxford, pp 105–125

Ma W-C (1972) Dynamics of feeding responses in *Pieris brassicae* Linn. as a function of chemosensory input: a behavioural, ultrastructural and electrophysiological study. Meded Landbhouwhogesch Wageningen 72-11:1–162

Ma W-C (1977) Alterations of chemoreceptor function in armyworm larvae (*Spodoptera exempta*) by a plant-derived sesquiterpenoid and by sulphydryl reagents. Physiol Entomol 2:199–207

Mitchell BK, Harrison GD (1985) Effects of *Solanum* glycoalkaloids on chemosensilla in the Colorado potato beetle. A mechanism of feeding deterrence. J Chem Ecol 11:73–831

Mitchell BK, Sutcliffe JF (1984) Sensory inhibition as a mechanism of feeding deterrence: effects of three alkaloids on leaf beetle feeding. Physiol Entomol 9:57–64

Mitter C, Futuyma DJ (1983) An evolutionary-genetic view of host-plant utilization by insects. In: Denno RF, McClure MS (eds) Variable Plants and Herbivores in Natural and Managed Systems. Academic Press, New York, pp 427–459

Pasteels J-CG, Rowell-Rahier M (1983) The chemical ecology of defense in arthropods. Annu Rev Entomol 28:263–290

Pluthero FG, Singh RJ (1984) Insect behavioral response to toxins: practical and evolutionary considerations. Can Entomol 116:57–68

Price PW (1983) Hypotheses on organization and evolution in herbivorous insect communities. In: Denno RF, McClure MS (eds) Variable Plants and Herbivores in Natural and Managed Systems. Academic Press, New York, pp 559–596

Price PW, Bouton CE, Gross P, McPheron BA, Thompson JN, Weis AE (1980) Interactions among three trophic levels: influence of plants on interactions between insect herbivores and natural enemies. Annu Rev Ecol Syst 11:41–65

Prokopy RJ, Averill AL, Cooley SS, Roitberg CA, Kallet C (1982) Variation in host acceptance pattern in apple maggot flies. In: Visser JH, Minks AK (eds) Proc 5th Int Symp Insect-Plant Relationships. Pudoc, Wageningen, pp 123–129

Pyke GH, Pulliam HR, Charnov EL (1977) Optimal Foraging: a selective review of theory and tests. Quart Rev Biol 52:137–154

Rathke BJ (1976) Competition and coexistence within a guild of herbivorous insects. Ecology 57:76–87

Rhoades DF (1983) Herbivore population dynamics and plant chemistry. In: Denno RF, McClure MS (eds) Variable Plants and Herbivores in Natural and Managed Systems. Academic Press, New York, pp 155–220

Rodriguez E, Levin DA (1976) Biochemical parallelisms of repellants and attractants in higher plants and arthropods. In: Wallace JW, Mansell RL (eds) Biochemical Interactions Between Plants and Insects. Plenum, New York, pp 214–270

Rosenthal GA, Janzen DH (eds) (1979) Herbivores: Their Interactions with Secondary Plant Metabolites. Academic Press, New York

Schoonhoven LM (1973) Plant recognition by lepidopterous larvae. Symp R Entomol Soc Lond 6:87–99

Schoonhoven LM (1981) Chemical mediators between plants and phytophagous insects In: Nordlund DA (ed) Semiochemicals: Their Role in Pest Control. John Wiley, New York, pp 31–50

Schoonhoven LM (1982) Biological aspects of antifeedants. Entomol Exp Appl 31:57–69

Schroeder LA (1976) Effect of food deprivation on the efficiency of utilization of dry matter, energy, and nitrogen by larvae of the cherry scallop moth, *Calocalpe undulata*. Ann Entomol Soc Am 69:55–58

Schultz JC (1983) Habitat selection and foraging tactics of caterpillars in heterogeneous trees. In: Denno RF, McClure MS (eds) Variable Plants and Herbivores in Natural and Managed Systems. Academic Press, New York, pp 61–90

Simmonds MSJ, Blaney WM (1984) Some neurophysiological effects of azadirachtin on lepidopterous larvae and their feeding response. In: Schmutterer H, Ascher KRS (eds) Natural Pesticides from the Neem Tree and Other Tropical Plants. Proc 2nd Int Neem Conference. Rauisch-Holzhausen Castle, Germany, pp 163–179

Strong DR, Lawton JH, Southwood TRE (1984) Insects on Plants. Harvard University Press

Swain T (1977) Secondary plant compounds as protective agents. Annu Rev Plant Physiol 42:255–302

Szentesi À. Bernays EA (1984) A study of behavioural habituation to a feeding deterrent in nymphs of *Schistocerca gregaria*. Physiol Entomol 9:329–340

Usher B, Feeny P (1983) Atypical secondary compounds in the family Crucifereae: tests for toxicity to *Pieris rapae,* an adapted crucifer-feeding insect. Entomol Exp Appl 34:257–262

van Drongelen W (1979) Contact chemoreception of host plant specific chemicals in larvae of various *Yponomeuta* species (Lepidoptera). J Comp Physiol 134:265–279

van Emden HF (1978) Insects and secondary plant substances—an alternative viewpoint with special reference to aphids. In: Harborne JB (ed) Biochemical Aspects of Plant and Animal Coevolution. Academic Press, New York, pp 309–323

Waldbauer GP (1962) The growth and reproduction of maxillectomized tobacco horn-worms feeding on normally rejected non-solanaceous plants. Entomol Exp Appl 5:147–158

Wasserman SS, Futuyma DJ (1981) Evolution of host-plant utilization in laboratory populations of the southern cowpea weevil, *Callosobruchus maculatus*. Evolution 35:605–617

Wiklund C (1982) Generalist versus specialist utilization of host plants among butterflies In: Visser JH, Minks AK (eds) Proc 5th Int Symp Insect-Plant Relationships. Pudoc, Wageningen, pp 181–192

Winstanley C, Blaney WM (1978) Chemosensory systems of locusts in relation to feeding. Entomol Exp Appl 24:750–758

Wood D (1982) The role of pheromones, kairomones and allomones in the host selection and colonization behavior of bark beetles. Annu Rev Entomol 27:411–446

Chapter 11

Speculations Concerning the Large White Butterfly (*Pieris brassicae* L.): Do the Females Assess the Number of Suitable Host Plants Present?

MIRIAM ROTHSCHILD*[+]

> I dreamed; and in my dream I was a butterfly
> Chuang Tzu
> (4th Century B.C.)

I believe I owe the editors of this volume an apology. I am sure they expected a chapter by me relating to the oviposition of butterflies, and, what is more, I agreed to provide one. But my flea-like mind jumped off at a tangent. Instead of a chapter I am going to lay a little bit of speculation at Vince Dethier's feet. In fact it is an act of homage, not of insubordinate second childhood.

In Oxford we used to say that when a lecturer became so incompetent that he failed to attract any students to his lectures, he sank into a resentful coma dignified by the title of research. That is a fate that awaits some of us. There are others, unfortunately, who retreat behind a shower of cold water directed toward their pupils. I would, any day, plump for the comatose nonlistener rather than the thrower of icy water. At the other end of the spectrum we find those very rare individuals who never lose the ability to produce new and stimulating ideas and—even more rarely—those who can evoke them in others. Vince Dethier has this amazing quality. When you talk to him and listen to him *you* begin to feel clever and at the end of the day he somehow creates a new idea for you. He is the antithesis of the thrower of cold water—he is, in fact, the distributor of rainbows.

The present hypothesis that the Large White can assess the abundance of its food plant will be more difficult to prove than the previous suggestion that female *Pieris brassicae,* by means of a combination of sight, smell, and touch, could "count" their eggs (Rothschild et al. 1975; Rothschild and Schoonhoven 1977; (Behan and Schoonhoven 1978; see also Prokopy 1975; Rausher 1979).

Various authors assert that butterflies that investigate the suitability of their

*Ashton Wold, Peterborough PE8 5LZ, U.K.
[+]This chapter was accepted for publication on April 15, 1985.

food by "drumming" (a bout of rapid tarsal tapping) or scratching the surface
of the leaves (Ilse 1937; see also Ma and Schoonhoven 1973) take off and resume
searching unless the plant drummed upon belongs to one of the larval host
species *"in which case oviposition usually follows"* (Feeny et al. 1983). This
is not true so far as the Large White butterfly is concerned. Immediate oviposition only very rarely follows initial drumming on a host plant (Table 11.1).
Even a gravid, well-fed female, denied access to such a plant for 8 days after
pairing, will usually drum, on an average, 14 different host plants or leaves
before laying (Table 11.2). It is evident that one such contact is insufficient to
satisfy her requirements for triggering oviposition.

This curious behavior of the Large White could be due to a variety of causes.
The female may be ultrasensitive to different deterrents or rare attractants found
on or in its host. Thus we know that slug-damaged leaves or those carrying
egg batches are often avoided. We also have evidence that supports the view
(Mitchell 1977; Feeny et al. 1983; Chew and Robbins 1984; personal observations) that her decision to lay may be influenced by leaf size, shape, color
(see also Kolb and Scherer 1982), texture, position, surface waxes, turgidity,
tough or shiny cuticle, crinkly growth, ground or surface water or dew, as well

Table 11.1 Responses of female Large White butterflies on encountering a host plant[a]

	Brassica oleracea: "Greyhound" Protected from pests Leaves undamaged	*Brassica oleracea:* "Greyhound" Not protected from pests Leaves heavily damaged[d]
Host plant		
Glucosinolates in leaves	0.51 μmol/g leaf	2.6–6.2 μmol/g leaf
Date	May 1984	August 1984
No. of females released	34	68
No. of ovipositions	18	42
Total contact drummings[b] before oviposition	210[c]	972[c]
Range of contact drummings before laying	3–35	2–80
Mean of contact drummings per butterfly	12	23
Less than five drummings before oviposition	3	6
No. of butterflies leaving cabbage patch without laying	16	26
Drumming contacts before leaving area	118	363
Mean of contact drummings before leaving	7	16

[a]Females were released in cabbage patch 2 days after pairing
[b]The term "drumming" refers to a bout or *series* of rapid tarsal tapping, not a count of single
tarsal tappings.
[c]Significant difference at the 5% level.
[d]A mean of 15 contact drummings before oviposition was recorded for 47 mated females
released in this area (August 1985).

Table 11.2. Number of drummings before first oviposition (125 females)

Time	No. of drummings (mean)
10–11 A.M.	19.07[a]
11–12 A.M.	16.32
12–1 P.M.	14.37
1–2 P.M.	17.10
2–3 P.M.	10.58[a]

Days since mating	No. of drummings (mean)
1	13.00
2	21.50
3	12.89
4	13.33
5	13.85
6	20.71
7	15.16
8	14.08

	No. of drummings (mean)
Temperature above 24°C[a] (with sun)	16.66
Temperature below 24°C[a] (dull sky)	10.60

[a]These is no significant difference in the number of drummings at 10–11 A.M. and 2–3 P.M. or in the number of drummings made at temperatures above and below 24°C.

as age, volatiles, and chemical constituents. The most powerful chemical deterrents in a plant containing glucosinolates are cardiac glycosides in the host tissues including petals (Rothschild et al. 1975). Recently Lundgren, Stenhagen, Alborn, and Karlsson (unpublished) have identified these substances in *Cheiranthus* and shown they are contact deterrents that completely override the glucosinolate cues present. The fact that different foliage on the same host plant contains different concentrations of glucosinolates complicates this issue (see Table 11.3). The proximity of aromatic or concealing vegetation, disturbances caused by hunting wasps or other ovipositing Lepidoptera (personal observations), and, of course, many aspects of weather and light conditions can also affect her choice. It is possible that this butterfly is genetically primed to keep "moving on"—the curious restlessness that assails her during oviposition adds to the chances of a wide distribution of her eggs. What triggers oviposition? May it not be the cumulative effect of the stimulation received through a series of bouts of contact drummings with leaves containing glucosinolates? Irrespective of the time spent in the cabbage patch, either actively investigating or basking, and irrespective of the collection of unsuitable hosts drummed upon

Table 11.3. Glucosinolate concentrations (μmol glucosinolate /g leaf) in host plants used in experiments[a]

Cabbage (*Brassicae oleracea*), Greyhound cultivar

Seedlings (May)	2.48 μmol
Young sprouts on old plants (May)	6.59 μmol
Mature leaves (May)	0.51 μmol
Young undamaged leaves (August)	3.50 μmol
Mature undamaged leaves (August)	2.60 μmol
Mature, heavily caterpillar-damaged leaves (August)	6.20 μmol

Rape (*Brassica napus*), spring sown seed, Jet Neuf cultivar

Line 80/92 leaves (May)	13.24 μmol
Line 80/129 leaves (May)	1.72 μmol
Line 80/92 leaves (Nov.)	8.90 μmol
Line 80/129 leaves (Nov.)	0.36 μmol

"Nasturtium" (*Tropaeolum majus*) Sutton's Gleam cultivar

Leaf (May)	0.65 μmol

[a] The estimates were made at the Plant Breeding Institute at Cambridge under the direction of K.F. Thompson.

previously, the number of such contacts on host plants before the first egg is laid is about 10–25. This ensures that there is adequate host material in the vicinity before the female lays, perhaps as many as 150 eggs per batch. It would be important for her to assess plant and foliage availability.

Materials and Methods

The Plants

Cabbage *(Brassica oleracea)* "Greyhound" Cultivar

Seedlings contained 2.48 μmol glucosinolates/g leaf, but this concentration fell as low as 0.51 μmol in mature leaves in May. The concentration in young sprouts growing from old plants later in the season rose sharply to 6.59 μmol glucosinolate/g leaf.

These plants were grown from seed, pricked out in trays, and then planted in the centre of the greenhouse to form a patch of five parallel rows of 16 plants each. A tiled path a meter wide separated this patch from a border of mixed flowering plants. When full grown, the lowest leaves of each cabbage just touched those of the adjacent plant. They were well watered daily.

The first spring planting was extraordinarily uniform by the first of May. The cabbages were large, all around the same size, and remarkably similar in growth, form, and color. They lacked young sprouts (present in some of the summer planting), and were screened from all garden pests, and consequently the leaves were undamaged by slugs, insect larvae, aphids, white fly, etc. From the end of May to the beginning of August the plants were not protected and were exposed to the full range of herbivores and pests. The plants, although all of the same age, had developed unevenly: some were very large and had flowered, whereas others remained relatively smaller. A few carried discolored leaves and many were considerably damaged by larvae of the Small White butterfly *(Pieris rapae)* as well as larvae of the Large White which had had access to the patch. No slugs or aphids were visible, but a fairly well-distributed infestation of white fly was present. The glucosinolate concentration was 2.6 μmol in old and undamaged leaves, 3.5 μmol in young leaves, and 6.2 μmol in old caterpillar-damaged leaves (Table 11.3).

Rape *(Brassica napus)*

Seed of a high glucosinolate strain (line 80/92 from Jet Neuf cultivar) and a low glucosinolate strain (line 80/129) were received from K. F. Thompson (Plant Breeding Institute, Cambridge). Both lines were obtained as naturally occurring haploids and the chromosome number doubled with colchicine to give homogeneous true-breeding lines. The glucosinolate content increased considerably when they were grown from spring-sown seed in adjacent plots at Ashton, but the difference between the two strains was maintained. The high strain then contained 13.24 μmol glucosinolate/g leaf and the low strain 1.72 μmol. They were not examined for carotenoid content. Unfortunately their growth rate was so variable that they could not be used for choice experiments, either as patches or even as single whole plants. Instead, matching branches from both strains were selected and exposed in glass jars (in the greenhouse) to ovipositing females. The containers were transposed periodically to avoid position effects.

"Nasturtium" *(Tropaeolum majus)* Cultivar "Sutton's Gleam"

This strain has been grown from seed at Ashton for 10 years. It had the highest carotenoid concentration in the leaves yet found by us in any plant (6756 μg/ g) (Rothschild et al. 1977), but glucosinolate concentrations were low, 0.65 μmol glucosinolate/g leaf.

A bucket of water containing cut flowers of *Buddleia* and *Senecio* was placed in the center of the cabbage patch.

The greenhouse in which the cabbage patches were planted was 30 meters long, 7 meters wide, and 5 meters high. It could be opened at both ends, and there were also several apertures in the glass roof. The tiled floor as well as the plants was watered every morning.

The Butterflies

Those used in the experiments were *Pieris brassicae* from Brian Gardiner's laboratory-bred Cambridge strain (David and Gardiner 1962). They were reared from egg to adult on cabbage, Greyhound cultivar, at Ashton. During the last (1984) series of experiments, wild-caught, free-flying Small White butterflies

(Pieris rapae) were introduced into the greenhouse to fly together with the Large Whites.

The Experiments

Female *P. brassicae* had unlimited access to varied plant nectar sources after eclosion. They had no access to suitable host plants until paired and set free in the greenhouse, but they drummed indefatigably on these nonhost plants. Releases were made singly at 10:30 a.m. 1–8 days after pairing. Each female was removed after the first oviposition and another specimen released. By this method habituation, conditioning, and other forms of learning were avoided. Furthermore the rearing of the larvae on the same food plants, the control of pairing and of time spent nectaring, the uniform development of the plants in the cabbage patch, and their screening against herbivores somewhat reduced the various confusing and conflicting factors that may influence oviposition. Since the age and history of the specimens were known, the fecundity of the females—a factor known to influence their oviposition behavior—could be allowed for, with reasonable confidence. *P. brassicae*, in the greenhouse, shows an egg-laying pattern of production rising to a maximum about the third day after pairing, although this is variable (Table 11.4), and also a well marked diurnal pattern, egg-laying being faster and the female less "choosy" between 10 a.m. and 1 p.m. irrespective of previous experience (Table 11.4). It is perhaps unnecessary to stress the characteristic individual variation and brood variation in every facet of butterfly behavior (Papaj and Rausher 1983), which is always present to confuse the complacent observer.

In order not to startle or disorientate the females by their sudden release, they were placed gently on *Buddleia* or *Senecio* flowers in the center of the cabbage patch. After a few moments feeding on the blooms they commenced a search for suitable host plants. They usually first settled on a cabbage in close proximity to the flowers. These females were then watched continuously and their movements recorded, until the first egg batch was laid. An alternative series of whole-day trials was made in which several females were released

Table 11.4. Number of eggs laid per hour by females hatched June 22. Host: Greyhound cabbage

Time eggs laid	No. of days after pairing					Grand total	Average
	2nd Temp 70° F	3rd Temp 90° F	4th Temp 90° F	5th Temp 80° F	6th Temp 80° F		
9–10 A.M.	22	17	200	nil	nil	239	48
10–11 A.M.	454	1749	595	498	280	3576	715
11–12 A.M.	684	849	505	196	178	2412	482
12–1 P.M.	316	272	183	159	67	997	199
1–2 P.M.	668	136	38	34	72	948	190
2–3 P.M.	nil	82	nil	5	nil	87	17
Total	2144	3105	1521	892	597	8259	

into the vicinity of glass jars containing carefully selected leaves of cabbage, rape, and nasturtium.

Results

The initial morning oviposition of the inexperienced Large White is characterized by the laying of relatively small egg batches and by her "choosiness." This is so whether or not she is at the beginning of the egg-laying period (David and Gardiner 1962; personal observations). During the first hour of egg-laying she reacts more strongly to such deterrents as egg batches on leaves and slug damage than she does later in the day when her decisions are arrived at a little more speedily (see Table 11.2). In the early morning she frequently interrupts her host searching to settle and bask on a leaf for 10 min or more. This time between her release and first oviposition may be as long as 1 hr 10 min.

The female on leaving the central flowers usually makes contact with the host plant in one of three ways:

1. Flies to a leaf, lands, and either remains resting with her wings closed, or opens her wings slightly, and, holding them apart, basks in the sunshine.
2. Flies to a leaf, lands, and drums on the surface with her forefeet, simultaneously fluttering her wings; very occasionally she drums without fluttering.
3. Flutters persistently against a vertical leaf without landing, but scrabbling the surface with her legs and touching it with her wings. She passes in this fashion from leaf to leaf. (Many cabbages in nature present only vertical or semivertical leaves to an ovipositing butterfly. Greyhound cultivar lacks the flat rosette type of growth.)

When females are about to oviposit, their flight is highly characteristic—very slow, almost feeble, and fluttering among the lower leaves of the plants (see also Lundgren 1975; Klijnstra 1985). If females are searching in a less persistent manner their flight is faster and stronger, and they tend to fly over plants rather than flutter among them near the ground. An observer can usually tell from the flight of the female whether she has decided the area is satisfactory and will settle down and lay, or whether she is uncertain and may eventually leave. Very often she will abandon the patch with a sweeping flight, but return again shortly after sampling nonhost plants nearby. There was no inclination in these returnees to lay on plants along the edge of the area as Jones (1977) reported for *P. rapae* in the field. They landed in any part of the cabbage patch.

However, 38–47% of those that left the cabbages to nectar, or search further afield, did not return. This was the case whether the butterflies had been paired 1 or 8 days previously. In the field this type of behavior makes observations very arduous if one attempts to follow individual specimens.

The female butterflies that drum on leaves are presumed to be testing their surface for suitability as host plants (Ilse 1937; Klijnstra 1982; Calvert and Hanson 1983; Chew and Robbins 1984). Unlike many nymphalids (see Dixon et al. 1978, plate 1a and b for photos of *Danaus plexippus* antennal host-testing), *P. brassicae* only rarely uses her antennae for examining oviposition sites. Although these organs are well supplied with olfactory sensilla (Behan and Schoonhoven 1978), and she taps the leaf surface with them at other times, she appears to rely almost exclusively on her tarsal receptors for initially testing the suitability of leaves for egg-laying and perhaps thus releasing volatiles from the cuticle. She passes from plant to plant either fluttering close to the surface or landing and drumming again. One female drummed on 35 plants before flying out of the area, not to return. The average number of preoviposition drumming series in the cabbage patches was between 12 and 23 (Table 11.1). When the female decides to lay, she settles near the edge of a leaf—obtaining the desired foothold on the shiny semivertical surface often appears difficult—snaps shut her wings, and curls her abdomen round to deposit her eggs on the underside.

Their initial reluctance to lay on what often appears to us to be perfect and readily available host plants was evident when females were released in the May cabbage patch which contained superb specimens that had been successfully shielded from all garden pests and herbivores. After the usual investigation of a selection of the plants a few more females than during the later trials—47% compared with 38%—left the area and did not return. Although the mean of individual drummings was somewhat lower, the basic pattern was similar (see Table 11.1). This was also the case if three suitable but different hosts, arranged in glass jars, were presented simultaneously to ovipositing females. Usually drummings were made on all three species of plants (see Table 11.5)

Table 11.5. Responses of 38 mated female Large Whites released together above three jars containing cabbage ("Greyhound"), rape (strain 80/92), and nasturtium ("Sutton's Gleam")[a]

Plant and glucosinolate concentration	Cabbage: 0.51 μmol/g leaf	Rape: 13.24 μmol/g leaf	Nasturtium: 0.65 μmol/g leaf
No. of ovipositions[b]	12	18	7
No. of bouts of drummings on each host	206	268	102
Total bouts of drummings on cabbage, rape, and nasturtium before ovipositing on one host	165	359	64
Mean number and (range) of bouts of drumming before ovipositing	14 (1–52)	20 (1–55)	9 (4–16)

[a]The behavior of *P. brassicae* vis-à-vis nasturtium is distinctive and not easy to explain. Although other chemically suitable plants are usually more attractive if offered together, it drums less before oviposition on this species (see also Table 11.6), but the difference here is not significant.

[b]Number of butterflies contacting all three hosts before laying was 18, two hosts before laying, 5, and one host before laying, 15.

and whether eggs were eventually laid on the cabbage, rape, or nasturtium, the number of preoviposition drummings was not significantly different (see also Table 11.6), although if any preference was shown it was for the rape.

On the whole, hot sunny weather resulted in more activity and slightly more preoviposition drummings, and cloudy, dull weather in less, but here again the differences were not significant (Table 11.2). Surprisingly there was no marked difference between females released 1 or 8 days after mating. Confined after copulation without a host plant, the Large White retains her eggs, and, unlike the Small White, only rarely lays on the "wrong" host or the walls of the breeding cage or the flower pot—behavior that is characteristic of many females deprived of a suitable host (Papaj and Rausher 1983, p 87). Yet, released on her foodplant, although her abdomen is bulging with eggs, the Large White goes through the routine search and the usual series of drummings. However one striking exception was recorded. Forty-one mated females deprived of host plants for 6 days after copulating were released on plants of rape that had just come into bloom. After nectaring on the flowers for 3–5 min the butterflies laid comparatively rapidly. The mean number of drummings prior to laying was 8, and 17 females drummed less than five times before ovipositing. Possibly nectaring on the host plant that contained glucosinolates provided a preparatory stimulus.

Careful watching revealed that occasionally a female, after 18–20 different drummings or scratching contacts with the host, would dart to the nearest cabbage and lay precipitately without any investigation at all and without the usual "positioning" maneuver near the edge of the leaf; this could result in a batch of eggs being placed on the upper surface. Feltwell (1982, p 66) noted that in the field in his experience 50% of eggs were laid on the upper surface, whereas in the laboratory in cages Klijnstra (1985) found that this rarely occurred. Furthermore the female may then lay immediately on a plant that she had rejected

Table 11.6. Responses of mated females released among selected leaves in glass jars (not tested together)

	Cabbage: (*Brassica oleracea*) "Greyhound" cultivar	Nasturtium: (*Tropaeolum majus*) "Sutton's Gleam"
Plant species		
Glucosinolate concentration in leaves	0.51 μmol/g leaf	0.65 μmol/g leaf
No. released	75	14
No. of ovipositions	75	14
No. of drumming bouts before laying	1196	167
Mean no. of drumming bouts	15.95	11.9[a]
Range of drumming bouts before laying	1–59	2–39
Less than five drumming bouts before oviposition	6	4

[a]Difference not significant.

and abandoned a few moments previously following a thorough contact-with-drumming appraisal.

Precipitate egg-laying without leaf and site testing was also noticed in the female Monarch butterfly ovipositing in the greenhouse. After laying an egg with elaborate care, scratching vigorously with the fore tarsi, followed by careful dipping and tapping with the antenna, and contact with the tip of the abdomen, the female darted to an adjacent host plant *(Asclepias curassavica)* growing at the same height, and laid another egg "recklessly" and in haste, on the same type of leaf, but without any investigation whatsoever.

These observations can be summarized thus:

1. The Large White does not lay immediately after finding a suitable larval host plant and drumming on the surface of its leaves. Usually a series of contacts on leaves containing glucosinolates occurs before oviposition.
2. It is difficult to assess the role of glucosinolate concentration in stimulating oviposition owing to the variation in different parts of the host plant. The evidence suggests, however, that although glucosinolates are a necessary cue for oviposition, their concentration plays no part in attracting the female or in cutting short serial drumming.
3. Plants protected from herbivores do not appear more attractive than unprotected plants, and serial drumming occurs on both.
4. The higher mean number of drummings on the unprotected plants may have been due to the August temperatures rather than the presence of more deterrents and a higher concentration of glucosinolates.
5. After about 18–20 plants have been tested by the female "precipitate" laying, i.e., without any drumming, can occur.

Discussion

There have been few studies made on the behavior of *P. brassicae* in the field, in sharp contrast to *P. rapae,* which has attracted a great deal of attention (for references see Ives 1978; Chew and Robbins 1984). These two butterflies, despite their superficial similarity, and the features they share with other members of the family Pieridae, are remarkably different, especially with regard to host finding and oviposition. Nevertheless, there is some evidence that repeated contacts with the host plant increase the Small White's responsiveness—unconnected with ovarian development (Traynier 1979).

The fact that the female Small White lays eggs singly and rapidly whereas the Large White lays slowly in batches has produced contrasting methods of host and site assessment, although in the garden they share similar host plants. Egg-batch ovipositors face certain specific risks not shared by single-egg ovipositors, particularly with regard to the available food supply, which must provide for about 70–150 larvae feeding up in close proximity on the same plant. Mitchell (1977) drew attention to the fact that in the field one very rarely finds larvae of more than one instar on a plant, which is not true of *P. rapae.*

It is important for both species to subscribe to the "egg spreading syndrome"

(Root and Kareiva 1984). This effect is achieved differently. *P. rapae* females' flight behavior results in the eggs of an individual being scattered over a large area. According to these authors the ovipositing Small White flies over an average of 5–10 host plants per single egg laid, and furthermore lays on only 30–50% of the apparently suitable hosts on which it lands. Its tendency to follow linear flight paths is an important contributing cause to the wide distribution of eggs. Jones (1977) showed that the accumulation of eggs on the peripheral plants in cabbage patches can be explained as a statistical result of this flight pattern. The Large White, on the other hand, once she has found a stand of acceptable plants, spreads her eggs by means of careful plant-host assessment at ground level, and appears far more attentive to detail. This mood only changes when she has drummed on a large number of different leaves and plants. Mitchell (1977) supports these observations for he concluded that initially the Large White displays a well-defined pattern of selection but that eventually "any plant can be selected as a host."

The role of the different glucosinolates in plant tissues as oviposition cues for White butterflies in not really understood. It is, of course, well known that the negative features of a dummy leaf are overcome if the surface is painted with sinigrin (David and Gardiner 1962) or cabbage leaf extract (Rothschild and Schoonhoven 1977) and that *P. brassicae* will then lay upon either or both. They will also lay if their feet are brushed with sinigrin. Furthermore Mitchell (1977) has demonstrated that in the field the ovipositing *P. brassicae* will select host plants of the wild cabbage *(Brassica oleracea* subsp. *oleracea)*, which showed a strongly positive reaction to the Guignard picrate test, and he suggested that allyl nitrile is a specific attractant for *P. brassicae*. Apparently *P. rapae* can respond differently, not only to the various glucosinolates, of which no fewer than 80 have been identified (Van Etten and Tookey 1979), but can discriminate among different combinations of these compounds in plants (Hovanitz and Chang 1963; Rodman and Chew 1980). From our own observations (see also Dixon et al. 1978) it would appear that the internal host plant environment may provide a different setting for the identical secondary plant substances, which consequently may provoke a different response in the butterflies.

P. brassicae shows a slight but not significant preference for rape foliage with a higher glucosinolate concentration than cabbage, when they are offered together. The butterflies drummed more frequently and tended to lay more egg batches on the rape leaves (see Table 11.5) although again the difference was not significant. On the other hand they selected cabbage leaves over the nasturtium leaves, both with regard to contacts and egg-laying, although the nasturtium contained a slightly higher concentration of glucosinolates. When butterflies were offered a choice of well-matched sprigs of the high and low glucosinolate strains, they sometimes selected one, sometimes the other, and offered the observer no clue as to the cause. Thus during five experiments (see Table 11.7), 91 batches (2921 eggs) were laid on the high- and 102 batches (3691 eggs) on the low-glucosinolate rape. In another series of 10 trials the reverse trend was demonstrated (see Table 11.7): 128 batches (3944 eggs) were laid on the high-, and 80 batches (2551 eggs) on the low-glucosinolate strain. Sometimes

Table 11.7. Oviposition of Large White on spring-sown Rape with high and low concentration of glucosinolates in foliage (plants tested November 1983)

May-June 1983 (five trials)	High glucosinolate (strain 80/92) 13.24 μmol/g leaf	Low glucosinolate (strain 80/129) 1.72 μmol/g leaf
Egg batches laid	128[a]	80[a]
Eggs laid	3944[a]	2551[a]
September-October 1983 (10 trials)		
Egg batches laid	91	102
Eggs laid	2921[a]	3691[a]

It is possible the glucosinolate content of the foliage altered during the summer, but it seems unlikely this could have induced the switch in the butterfly's choice. Although the glucosinolate content was lower in autumn-sown plants, the difference between the two strains was maintained (i.e., 0.36 μmol/g for strain 80/129 and 8.90 μmol/g for strain 80/92).
[a]Significantly different at 1% level.

the butterflies apparently favored the high strain in the morning and the low strain in the afternoon. The key is evidently the glucosinolates, but the lock can be surprisingly complex. The butterflies know the combination. At present we do not.

Nor do we understand the relative importance of specific deterrents to the ovipositing *P. brassicae*. In the greenhouse, for instance, when there are no other alternatives, the female selects a "clean" leaf to one bearing conspecific eggs. But if a slug-damaged leaf is offered with an egg-laden leaf as an alternative, she will usually lay on the latter. But these are relatively minor mysteries. The most puzzling and striking feature of the oviposition of the female Large White is her initial reluctance to lay on apparently eminently suitable host plants when they are offered to her, and her compulsive testing of so many alternatives. As we have seen, this is equally true of a recently mated female or one deprived of host plants for 8 days after copulation. There are very few cases of immediate egg "dumping," described for similarly deprived Small Whites (Root and Kareiva 1984). The Large White's behavior suggests that one contact-with-drumming per host plant is usually insufficient to convey the message and to release egg-laying—a series of such contacts is required. Eventually the accumulated stimuli imperiously override the plethora of negative cues that apparently assail the Large White and drive her remorselessly on to the next and the next plant. This would explain the curious phenomenon, following multiple drumming, of "reckless" laying. For if the stand of food plants is large enough, the future larvae will be able to find sufficient food close by, even if their original host plant is completely consumed: in such circumstances careful testing ceases to be necessary and the butterfly gets the "green light" to lay and can safely conserve her energy for the morrow.

Here I will end my little piece of speculation in honor of Vince Dethier, but I would like to terminate this brief chapter on a different and perhaps more serious note—for this is a special occasion.

Vince Dethier is an original and intuitive thinker and he is also a perceptive, hard-headed, down-to-earth analytical scientist. But as a teacher he is more than the sum of these qualities for he possesses the golden touch—he inspires his pupils with a sense of the romance in the insect world, so that they see things in a rosy light; the little events of this miniature cosmos strike us in a more imaginative way, and give rise to felicitous ideas. He takes the dullness and boredom out of scientific method.

A few illustrations show, for me, a romantic slant on the White Butterfly. A section through the cuticle of a pupa killed in the act of matching its background (Figure 11.1)[1] (stained with Heidenhain's azan stain) shows you three layers of the cuticle. The outermost and densest layer does not admit any stain at all and appears golden brown in color. The second layer admits azocarmine and stains a brilliant red, whereas the third layer takes up the stain of aniline blue. This, to me, is an object as beautiful as a modern picture by Paul Jenkins or Georgia O'Keefe.

You will recall situations in the lives of various larvae and pupae when the cuticle, especially of tubercles, excludes certain colors and admits others, usually under hormonal control. Thus the lateral tubercles of certain silk moth larvae are bright blue with pterobilin in the tissues (Figure 11.2), whereas in other species the dorsal tubercles are pillar-box red and buttercup yellow, due to the presence of carotenoids—pterobilins being excluded. The body is green, for here, in the cuticle and underlying tissues, pterobilins and carotenoids are both present and mix in the eye of the beholder to produce this cryptic colouration. Some silk moth caterpillars change the color of their tubercles at different molts. The switching to and fro of these vivid tints is something that never ceases to enthral and puzzle the naturalist.

> Soon spreads the dismal shade
> Of Mystery over his head;
> And the Catterpillar and Fly
> Feed on the Mystery.[2]

A different angle to entomological romance is the hungry caterpillar of *Pieris* eating—not Blake's mystery, but the wings of its moribund mother. To the onlooker this act of matricide revealed immediately that the imago stores the glucosinolates sequestered by the larva (Aplin et al. 1975).

Another example is of a male Large White caught by the tongue in the pollen trap of *Arauja* (an Asclepiad plant) and in this hapless situation, where he cannot defend himself against the outrage, he is sexually assaulted by another passing male.

> The small gilded fly doth lecher in our sight.[3]

A far grimmer sidelight is thrown upon the large White by reason of the extremely toxic protein-like substance (pierin) found in both larva and pupa (Dempster 1967). Marsh and I injected the white (laboratory) mouse with an

---

[1]Figures 11.1 to 11.3 appear on page 181.
[2]William Blake, "Human Abstract."
[3]William Shakespeare, *King Lear:* IV, 12.

extract of the pupa, which caused massive hemorrhages into its tissues and death within 10–15 hr (Marsh and Rothschild, 1974). Today I would not think such an experiment justified the suffering inflicted on an animal, but in those far off times I kidded myself into believing that my work might be of medical significance. Fortunately attitudes are changing.

One of the most delightful sights in the garden is Large Whites massing on the flowers of heliotrope. With poetic license, Keats, and many others, declared butterflies were rose lovers,

> A butterfly with golden wings broadparted,
> Nestling a rose, convulsed as though it smarted
> With pleasure.[4]

Roses are basically beetle flowers and the poets missed the frantic delight engendered by cherry pie. Does the nectar of heliotrope contain pyrrolizidine alkaloids—like various strains of ragwort? I have seen the Large White imbibing the exudate from the foliage of *Passiflora*, which contains HCN or its precursors (Figure 11.3). Does the butterfly swipe such substances, after emergence, for its own defense purposes or for sexual stimulation? This is the sort of daydream one may indulge in while waiting in a Monday morning traffic jam. Which brings me back to my central theme—the art of persuading one's students to daydream—an art practised, maybe unconsciously, but with consummate skill by Vince Dethier. Because of course science, even the study of the angelic butterfly, becomes dull, arithmetical, and brittle without a touch of poetry.

> This dull chrysalis cracks into shining wings.[5]

Figure 11.4 is a reproduction of Van Gogh's famous picture of the prison yard. Here we are shown poor humanity trudging round the exercise ring—confined within the towering prison walls, or perhaps laboratory walls for all I know. But the arrow (my arrow, of course, not Van Gogh's) points to a "discovery," for no art critic seems to have spotted two White butterflies twirling in freedom and winged delight. For me they are the symbol of daydreaming—the poetry that Vince Dethier insinuates so cunningly into our factual information and knowledge. For the gift, of these special white butterflies —along with all your official and unofficial students, past, present, and future—Vince Dethier, I tender you my most heartfelt and grateful thanks.

Acknowledgments First and foremost I would like to thank Dorothy Lucas for many hours of careful observations and various constructive suggestions and calculations. I am deeply grateful to Kenneth Thompson and his staff, not only for providing the different strains of rape, but for testing plant material for glucosinolate content, despite an immense pressure of work. Brian Gardiner kindly sent me a constant supply of healthy butterflies and Bob Ford assisted with the preliminary work for this study in 1975. I would also

---

[4]John Keats, "Sleep and Poetry."

[5]Alfred Tennyson, "St. Simeon Stylites."

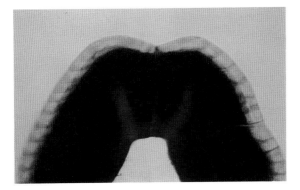

Figure 11.1. Section through the cuticle of the pupal *Pieris brassicae* stained with Heidenhain's azan.

Figure 11.2. Larval silkmoth, *Antheraea pernyi,* showing pterobilin concentrated in the tubercles.

Figure 11.3. *Pieris brassicae* imbibing exudate from leaf of *Passiflora*.

Figure 11.4. Vincent van Gogh; "The Prison Yard." The arrow points to two white butterflies. (Original in Pushkin Museum, Moscow.)

like to record my debt to Mike Singer, who read the typescript and offered invaluable criticism.

References

Aplin RT, Ward R d'Arcy, Rothschild M (1975) Examination of the large white and small white butterflies (*Pieris* spp.) for the presence of mustard oils and mustard oil glycosides. J Entomol Ser A 50:73–78

Behan M, Schoonhoven LM (1978) Chemoreception of an oviposition deterrent associated with eggs in *Pieris brassicae*. Entomol Exp Appl 24:163–179

Calvert WH, Hanson FE (1983) The role of sensory structures and preoviposition behavior in oviposition by the patch butterfly, *Chylosyne lacinia*. Entomol Exp Appl 33:179–187

Chew FS, Robbins RK (1984) Egg-laying in Butterflies. In: Ackery PR, Vane-Wright RI (eds) Biology of Butterflies. Academic Press, London, pp 65–79

David WAL, Gardiner BOC (1962) Oviposition and the hatching of the eggs of *Pieris brassicae* (L.) in a laboratory culture. Bull Entomol Res 53:91–109

Dempster JP (1967) The control of *Pieris rapae* with DDT. (I) The natural mortality of the young stages of *Pieris*. J Ecol 4:485–500

Dixon CA, Erickson JM, Kellett DN, Rothschild M (1978) Some adaptations between *Danaus plexippus* and its food plant, with notes on *Danaus chrysippus* and *Euploea core* (Insecta: Lepidoptera). J Zool Lond 185:437–467

Feeny P, Rosenberry L, Carter M (1983) Chemical aspects of oviposition behavior in butterflies. In: Ahmad S (ed) Herbivorous Insects. Academic Press, New York, pp 27–76

Feltwell J (1982) The large white butterfly: biology, biochemistry and physiology of *Pieris brassicae* (Linnaeus). Junk, The Hague

Hovanitz W, Chang VCS (1963) Ovipositional preference tests with *Pieris*. J Res Lepid 2:185–200

Ilse D (1937) New observations on responses to colours in egg-laying butterflies. Nature Lond 140:544–545

Ives PM (1978) How discriminating are cabbage butterflies? Aust J Ecol 3:261–276

Jones RE (1977) Movement patterns and egg distribution in cabbage butterflies. J Anim Ecol 46:195–212

Klijnstra JW (1982) Perception of the oviposition deterrent pheromone in *Pieris brassicae*. In: Visser JH, Minks AK (eds) Proc 5th Int Symp Insect-Plant Relationships. Pudoc, Wageningen, pp 145–151

Klijnstra JW (1985) Oviposition behaviour as influenced by the oviposition deterring pheromone in the large white butterfly, *Pieris brassicae*. Thesis, Wageningen

Kolb G, Scherer C (1982) Experiments on wavelength specific behavior of *Pieris brassicae* L. during drumming and egg-laying. J Comp Physiol A 149:325–332

Lundgren L (1975) Natural plant chemicals acting as oviposition deterrents on cabbage butterflies (*Pieris brassicae* L., *P. rapae* L. and *P. napi* L.). Zool Scr 4:253–258

Ma W-C, Schoonhoven LM (1973) Tarsal contact chemosensory hairs of the large white butterfly *Pieris brassicae* and their possible role in oviposition behaviour. Entomol Exp Appl 16:343–357

Marsh N, Rothschild M (1974) Aposematic and cryptic Lepidoptera tested on the mouse. J Zool Lond 174:89–122

Mitchell ND (1977) Differential host selection by *Pieris brassicae* (the large white butterfly) on *Brassica oleracea* subsp. *oleracea* (the wild cabbage). Entomol Exp Appl 22:208–219

Papaj DR, Rausher MD (1983) Individual variation in host location by phytophagous insects. In: Ahmad S (ed) Herbivorous Insects. Academic Press, New York, pp 77–124

Prokopy RJ (1975) Oviposition-deterring fruit marking pheromone in *Rhagoletis fausta*. Environ Entomol 4:298–300

Rausher MD (1979) Egg recognition: its advantage to a butterfly. Anim Behav 27:1034–1040

Rodman JE, Chew FS (1980) Phytochemical correlates of herbivory in a community of native and naturalized Crucifereae. Biochem Syst Ecol 8:43–50

Root RB, Kareiva PM (1984) The search for resources by cabbage butterflies (*Pieris rapae*): ecological consequences and adaptive significance of markovian movements in a patchy environment. Ecology 65:147–165

Rothschild M, Schoonhoven LM (1977) Assessment of egg load by *Pieris brassicae* (Lepidoptera: Pieridae). Nature 266:352–355

Rothschild M, Gardiner BOC, Valadon G, Mummery R (1975) The large white butterfly: oviposition cues, carotenoids and changes of colour. Proc R Entomol Soc Lond C 40:13

Rothschild M, Valadon G, Mummery R (1977) Carotenoids of the large white butterfly *(Pieris brassicae)* and the small white butterfly *(Pieris rapae)*. J Zool Lond 181:323–339

Traynier RMM (1979) Long-term changes in the oviposition behavior of the cabbage butterfly, *Pieris rapae,* induced by contact with plants. Physiol Entomol 4:87–96

Van Etten CH, Tookey HL (1979) Chemistry and biological effects of glucosinolates. In: Rosenthal GA, Janzen DH (eds) Herbivores: Their Interaction with Secondary Plant Metabolites. Academic Press, New York, pp 471–500

Author Index

Page numbers in brackets indicate references

Subject Index